Low-cost Smart Antennas

Microwave and Wireless Technologies Series

Series Editor, Professor Steven (Shichang) Gao, Chair of RF and Microwave Engineering, and the Director of Postgraduate Research at School of Engineering and Digital Arts, University of Kent, UK.

Microwave and wireless industries have experienced significant development during recent decades. New developments such as 5G mobile communications, broadband satellite communications, high-resolution earth observation, the Internet of Things, the Internet of Space, THz technologies, wearable electronics, 3D printing, autonomous driving, artificial intelligence etc. will enable more innovations in microwave and wireless technologies. The Microwave and Wireless Technologies Book Series aims to publish a number of high-quality books covering topics of areas of antenna theory and technologies, radio propagation, radio frequency, microwave, millimetre-wave and THz devices, circuits and systems, electromagnetic field theory and engineering, electromagnetic compatibility, photonics devices, circuits and systems, microwave photonics, new materials for applications into microwave and photonics, new manufacturing technologies for microwave and photonics applications, wireless systems and networks.

Low-cost Smart Antennas

Qi Luo, and Steven (Shichang) Gao
University of Kent, Canterbury, UK

Wei Liu
University of Sheffield, Sheffield, UK

Chao Gu
University of Kent, Canterbury, UK

Registered Offices
John Wiley & Sons, Inc., 111 River Street, Hoboken, NJ 07030, USA
John Wiley & Sons Ltd, The Atrium, Southern Gate, Chichester, West Sussex, PO19 8SQ, UK

Editorial Office
The Atrium, Southern Gate, Chichester, West Sussex, PO19 8SQ, UK

For details of our global editorial offices, customer services, and more information about Wiley products visit us at www.wiley.com.

Wiley also publishes its books in a variety of electronic formats and by print-on-demand. Some content that appears in standard print versions of this book may not be available in other formats.

Library of Congress Cataloging-in-Publication Data

Names: Luo, Qi, 1982- author. | Gao, Steven (Shichang), author. | Liu, Wei, 1974- author. |
   Gu, Chao, 1986 Sept. 24- author.
Title: Low-cost smart antennas / Qi Luo, Steven (Shichang) Gao, Wei Liu, and Chao Gu.
Description: First edition. | Hoboken, NJ : Wiley, [2019] | Series:
   Microwave and wireless technologies series | Includes bibliographical
   references and index. |
Identifiers: LCCN 2018042667 (print) | LCCN 2018050919 (ebook) | ISBN
   9781119422792 (Adobe PDF) | ISBN 9781119422877 (ePub) | ISBN 9781119422778
   (hardcover)
Subjects: LCSH: Adaptive antennas.
Classification: LCC TK7871.67.A33 (ebook) | LCC TK7871.67.A33 L86 2019
   (print) | DDC 621.3841/35–dc23
LC record available at https://lccn.loc.gov/2018042667

Cover Design: Wiley
Cover Images: Abstract texture © AboutnuyLove/iStock.com, Vector mockup © ExtraDryRain/iStock.com, SUV © Henrik5000/iStock.com, Airplane © lvcandy/iStock.com, Broadcast satellite © Madmaxer/iStock.com, Cellular camera © Flamell/iStock.com, High-speed train © Nerthuz/iStock.com, Low-cost smart antenna courtesy of the authors

Set in 10/12pt WarnockPro by SPi Global, Chennai, India
Printed in Singapore by C.O.S. Printers Pte Ltd

10 9 8 7 6 5 4 3 2 1

Contents

Preface

Smart antennas are antennas with smart signal-processing algorithms that can electronically reconfigure radiation patterns so that the maximum radiation is formed towards the desired directions while nulls are formed towards interfering sources. It is a key technology for many wireless systems, such as satellite communications, terrestrial mobile communications, inter-satellite links, radio-frequency identification, wireless power transmission, wireless local area networks, global navigation satellite systems, radars, remote sensing, and direct broadcast satellite television reception systems. Traditional smart antennas using phased arrays or digital beamforming adaptive arrays are rather complicated in structure, bulky, power hungry, and costly. For commercial applications, it is important to reduce the size, mass, power consumption, and cost of smart antennas. Recent decades have seen lots of progress in research and development in the field of low-cost smart antennas. It is foreseen that low-cost smart antennas will be widely implemented in the smart city, fifth-generation and future generations of mobile communications, smart homes, satellite communication on the move, the Internet of Things, the Internet of Space, and autonomous vehicles.

So far, most books on smart antennas have mainly focused on signal-processing algorithms, and there are few books specialising in antennas and the radio frequency (RF) hardware of smart antennas. The purpose of this book is to address practical antenna design and RF engineering issues in low-cost smart antennas by presenting various techniques for designing and implementing low-cost smart antennas. These techniques include the electronically steerable parasitic array radiator, the reconfigurable frequency selective surface, pattern-reconfigurable reflectarrays and transmitarrays, compact multiple-input multiple-output antenna systems, and the use of low-cost beamforming networks. Each topic is addressed with both theoretical explanations and practical design examples. Each chapter contains basic principles, design techniques, a detailed review of state-of-the-art development, and practical case studies to illustrate how to design low-cost smart antennas step by step. To provide readers with some basics of beamforming algorithms and their applications in smart antennas, Chapter 2 discusses the basic principles of beamforming and introduces some representative beamforming methods and algorithms for smart antennas. A review of the particular area of low-cost adaptive beamforming is also presented in this chapter, including hybrid beamforming and robust adaptive beamforming.

This book contains fundamental theory, many practical design examples, advanced design techniques, and case studies, thus it is a useful reference for people from both industry and academia who are interested in smart antennas. The references listed in each chapter offer additional sources of data for readers.

Acknowledgement

The authors would like to thank Sandra Grayson, Louis Manohar, and Kanchana Kathirvelu of Wiley for their help and guidance during the preparation of this book.

Dr Qi Luo would like to express his appreciation and gratitude to his family for their encouragement, understanding, and patience during the writing of this book. He also would like to express his thanks for technical discussions to the members of the antenna research group at the University of Kent, UK.

Professor Steven (Shichang) Gao would like to thank his wife Jun Li and his daughter Karen Yu Gao for their great understanding and support during the period of book writing. He also would like to thank all of his current and former students and research collaborators who contributed to the research work on low-cost smart antennas. In particular, thanks to Dr Haitao Liu, Dr Long Zhang, Mr Hang Xu, Dr Fan Qin, Mr Mingtao Zhang, Dr Benito Sanz, Professor Ted Parker, Dr Hanyang Wang, Dr Hai Zhou, Professor Xuexia Yang, Professor Yingzeng Yin, Professor Yongchang Jiao, Professor Ying Liu, Associate Professor Jianzhou Li, Professor Gao Wei, Professor Jiadong Xu, Professor Luigi Boccia, and Professor Amendola Giandomenico who made important contributions into the research on low-cost smart antennas.

Dr Wei Liu would like to thank all former and current members of his research group for their hard work and creativity, and especially those who joined his group at the early stage of his career. In particular, some of the work presented in the book is closely related to the research carried out by Lei Zhang, Qiu Bo, and Craig Miller.

Dr Chao Gu would like to thank Simon Jakes for the help he provided in prototyping the antennas. He also wishes to express his heartfelt gratitude to his wife, Dr Lu Bai, for her support and unwavering belief.

Parts of the research work presented in this book were supported by EPSRC grants EP/N032497/1, EP/P015840/1, and EP/S005625/1.

List of Abbreviations

2D-FFT	Two-dimensional fast Fourier transform
ABF	Analogue beamforming
AC	Alternating current
A/D	Analogue/digital
ADC	Analogue-to-digital converter
AF	Array factor
AFR	Array-fed reflector
AFSS	Active frequency selective surface
AMC	Artificial magnetic conductor
AOA	Angle of arrival
AR	Axial ratio
AWGN	Additive white Gaussian noise
BFN	Beamforming network
BP	Beam pattern
CAFSS	Cylindrical active frequency selective surface
CCC	Cross-correlation coefficient
CMA	Constant modulus algorithm
CP	Circularly polarised
DAC	Digital-to-analogue converter
dB	Decibels
DBF	Digital beamforming
DC	Direct current
DDC	Digital down-converter
DGS	Defect ground system
DOA	Direction of arrival
DSP	Digital signal processor
EBG	Electromagnetic band gap
ECC	Envelope correlation coefficient
ECM	Equivalent circuit method
EM	Electromagnetic
ESPAR	Electronically steerable parasitic array radiator
FBR	Front-to-back ratio
FDTD	finite difference time domain
FEM	finite element method
FIR	Finite impulse response

FM-ESPAR	Folded monopole ESPAR
FOM	Figures of merit
FP	Fabry–Perot
FPGA	Field-programmable gate array
FSK	Frequency shift keying
FSS	Frequency selective surface
GCPW	Grounded coplanar waveguide
GNSS	Global navigation satellite system
GPIO	General purpose input/output
GPS	Global positioning system
HIS	High impedance surface
HPBW	Half-power beamwidth
IF	Intermediate frequency
ILA	Inverted-L antenna
INR	Interference-to-noise ratio
LAN	Local-area network
LC	Inductor-capacitor
LCMV	Linearly constrained minimum variance
LCP	Liquid crystal polymer
LHCP	Left-hand circularly polarised
LMS	Least mean squares
LO	Local oscillator
LQI	Link quality indicator
LS	Least squares
LTCC	Low temperature co-fired ceramic
LTE	Long-term evolution
MCU	Micro control unit
MEMS	Microelectromechanical systems
MIMO	Multiple-input multiple-output
mm-wave	Millimetre-wave
MPR	Metamaterial polarisation-rotator
MSR	Mainlobe to sidelobe ratio
MTM	Metamaterial
MUSIC	MUltiple SIgnal Classification
NMSE	Normalised mean square error
PCB	Printed circuit board
PBG	Photonic bandgap
PER	Package error rate
PFGA	Field-programmable gate array
PIFA	Planar inverted-F antenna
PRS	Partially reflective surface
PSK	Phase shift keying
QPSK	Quadrature phase shift keying
RF	Radio frequency
RHCP	Right-hand circularly polarised
RLC	Resistor-inductor-capacitor
RLS	Recursive least squares

SAR	Specific absorption rate
SDL	Sensor delay line
SINR	Signal to interference plus noise ratio
SIW	Substrate-integrated waveguide
SLL	Sidelobe level
SMA	SubMiniature version A
SNR	Signal-to-noise ratio
SOI	Signal of interest
SP3T	Single pole triple throw
SRF	Self-resonant frequency
SRR	Split-ring resonator
TA	Transmitarray
TARC	Total active reflection coefficient
TDL	Tapped delay line
TE	Transverse electric
TM	Transverse magnetic
T/R	Transmit/receive
TTL	Transistor–transistor logic
ULAs	Uniform linear arrays
USB	Universal serial bus
VCO	Voltage-controlled oscillator
VNA	Vector network analyser
VSWR	Voltage standing wave ratio
WiMAX	Worldwide interoperability for microwave access
WLAN	Wireless local-area network

1

Introduction to Smart Antennas

1.1 Introduction

Smart antennas, also known as intelligent antennas or adaptive arrays, are a key technology for advanced wireless systems, such as satellite communications, inter-satellite links, radars, sensors, mobile communications (5G and beyond), wireless local area networks, global navigation satellite systems, and wireless power transfer. One of the most important features of smart antennas is electronic beam scanning or switching. Smart antennas enable wireless systems to achieve optimum performance and increase channel throughput by electronically steering maximum radiation towards the desired directions while forming nulls against interfering sources.

The adaptive array is 'smart' because it has signal processing units with smart signal processing algorithms. Recent years have seen the development of efficient algorithms for direction of arrival (DOA) estimation and adaptive beamforming. Algorithms for adaptive beamforming include the classical least mean squares (LMS) type algorithm, constant modulus algorithm (CMA) etc. Traditional smart antennas are, however, complicated in structure, bulky in size, and costly. Thus, it is highly desirable to reduce the size, mass, power consumption, and cost of smart antennas.

Generally speaking, smart antennas can be divided into three components: the antenna, the beamforming network (BFN), and the signal processing unit. It is believed that radio frequency (RF) designs such as the architecture design and configurations of antenna and BFN play an important role in determining the overall cost of a smart antenna. A good example is the phased array in which each of the antenna elements has its own RF chain, and the number of active antenna elements determines the number of transmit/receive (T/R) modules and the complexity of the BFN. An active phased array with 1000 antenna elements typically requires 1000 RF phase shifters, 1000 RF transceivers, and a highly complicated BFN, resulting in large size, heavy weight, high power consumption, and high cost. For civilian applications, it is crucial to develop low-cost smart antennas. Here low-cost smart antennas refer to antenna systems which can achieve electronic beam scanning, multiple beams or electronic beam-switching, and have significantly lower cost compared to traditional smart antennas such as phased arrays or digital beamforming smart antennas. Low-cost smart antennas can be achieved by designing innovative antenna system architectures that require a

Low-cost Smart Antennas, First Edition. Qi Luo, Steven (Shichang) Gao, Wei Liu, and Chao Gu.
© 2019 John Wiley & Sons Ltd. Published 2019 by John Wiley & Sons Ltd.

significantly reduced number of T/R modules and RF phase shifters, or simplified BFNs with low-cost beamforming algorithms.

This book focuses on the RF design of smart antennas from the aspect of the array antenna, BFN, and related beamforming algorithms. The main purpose of the book is to present the techniques of RF designs of low-cost smart antennas as well as the hardware implementations of the antenna and the BFN. As multiple-input multiple-output (MIMO) antennas are often regarded as one type of smart antennas, compact-size MIMO antennas are included as one special type of low-cost smart antennas in this book. Due to the importance of beamforming algorithms for smart antennas, one chapter on beamforming algorithms is also included and many examples are discussed.

This chapter will provide an introduction to the fundamental concepts of antennas, array antennas, and smart antennas, laying a foundation for the following chapters. Configurations of smart antennas are also explained and discussed.

1.2 Antenna Fundamentals

In this section, some fundamental parameters of antennas are briefly presented, including input impedance, bandwidth, radiation pattern, polarisation, efficiency, and gain. These are key parameters for an antenna and are critical for the radiation performance of smart antennas.

1.2.1 Antenna Impedance and Bandwidth

The input impedance of the antenna is defined as the ratio of voltage to current at the terminal of the antenna. It is the ratio of the voltage to current or the ratio of the appropriate components of the electric to magnetic fields at the feed point [1]. The impedance of the antenna is usually a complex number and it is frequency dependent. It can be expressed as

$$Z_A = R_A + jX_A \tag{1.1}$$

where Z_A, R_A and X_A represent the antenna impedance, antenna resistance, and antenna reactance at the terminal of the antenna, respectively. The antenna resistance includes the radiation resistance (R_r) and the loss resistance (R_L) of the antenna

$$R_A = R_r + R_L \tag{1.2}$$

The radiation resistance is related to the power radiated by the antenna, and the loss resistance is associated with the power dissipated in the antenna due to the losses from the dielectric material and conductor. For a multi-port antenna, as a result of the mutual impedance between different ports, the input impedance of the antenna becomes [2]

$$Z_{A,i} = V_i/I_i = Z_{ii} + \sum_{j \neq i} Z_{ij}I_j/I_i \tag{1.3}$$

where $Z_{A,i}$ represents the input impedance at port i, Z_{ii} is the self-impedance of the i_{th} port, Z_{ij} is the mutual impedance between ports i and j, and I represents the current at the port of the antenna. As shown in Equation 1.3, the input impedance at port i is related to the excitations from other ports through the mutual impedance. Ideally, if

Figure 1.1 The equivalent circuit of the input impedance of the antenna with transmission line.

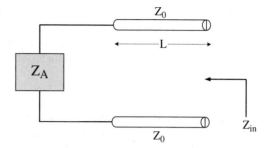

the mutual impedance is very small, the input impedance of each port is independent of the excitations of other ports. It is required that the antenna is impedance matched to the transmission line otherwise the antenna cannot radiate efficiently. As shown in Figure 1.1, when the antenna is terminated with a transmission with the impedance Z_0 and length L, the input impedance is

$$Z_{in} = Z_0 \frac{Z_A + jZ_0 \tan(L/\lambda)}{Z_0 + jZ_A \tan(L/\lambda)} \tag{1.4}$$

where λ is the wavelength in free space.

Figure 1.2 shows the input impedance of a typical probe-fed rectangular patch antenna. This patch has resonance at 9.75 GHz and the feeding coaxial cable has impedance of 50 Ω. This patch is printed on a 1.57 mm thick RT/Duroid 5880 substrate ($\varepsilon_r = 2.2$). As shown in Figure 1.2, at the resonance of the antenna the imaginary part of the input impedance is close to zero while the real part of the input impedance is close to 50 Ω. The reflection coefficient of the antenna is defined as

$$\Gamma = \frac{Z_{in} - Z_0}{Z_{in} + Z_0} \tag{1.5}$$

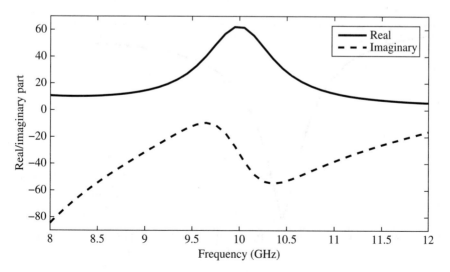

Figure 1.2 The input impedance of a typical probe-fed square patch antenna.

The concept of voltage standing wave ratio (VSWR) is introduced as a measure to show how well the antenna is matched. It is defined as the ratio of the maximum voltage (V_{max}) to the minimum voltage (V_{min}) in standing wave pattern along the transmission line. It is related to the reflection coefficient by

$$VSWR = \frac{V_{max}}{V_{min}} = \frac{1 + |\Gamma|}{1 - |\Gamma|} \qquad (1.6)$$

As shown in Equation 1.6, VSWR is a real number that is always greater than or equal to 1. A VSWR of 1 indicates that there is no mismatch loss, while higher values of VSWR imply that there is large mismatch loss. Another parameter that can be used to quantise the matching of the antenna is return loss, which is defined as the ratio of rejected power against the input power to the antenna port. It is specified in decibels (dB) and is expressed as

$$RL = -20 \log |\Gamma| = -20 \log \left(\frac{VSWR - 1}{VSWR + 1} \right) \qquad (1.7)$$

Another parameter that is equivalent to the return loss is the amplitude of the reflection coefficient |S11|. The |S11| represents how much power is reflected from the antenna. Generally speaking, the bandwidth of the antenna is defined as the frequency range where the return loss is larger than 10 dB (|S11| < −10 dB) or the VSWR is smaller than 2. In some applications, such as the mobile phone devices, the bandwidth of the antenna is defined as return loss larger than 6 dB while in base station application it is always desirable to have the return loss larger than 15 dB. Figure 1.3 shows the return loss and VSWR of the patch antenna presented in Figure 1.2. The 10 dB return loss bandwidth of the patch is approximately 5.8% at the central frequency of 9.75 GHz.

1.2.2 Radiation Patterns and Efficiency

The radiation pattern of the antenna shows the distribution of the radiated power in the far-field. It is defined as 'a mathematical function or a graphical representation of the

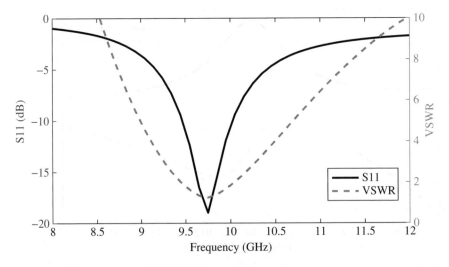

Figure 1.3 The return loss and VSWR of the patch antenna.

radiation properties of the antenna as a function of space coordinates' [1]. The power varies as a function of the angles that are observed in the far-field region of the antenna. In the far-field region, the radiation pattern does not change with distance. The far-field is defined as

$$R > \frac{2D^2}{\lambda} \tag{1.8}$$

$$R \gg \lambda \tag{1.9}$$

$$R \gg D \tag{1.10}$$

where D is the maximum dimension of the antenna and λ is the free space wavelength. The reactive near-field is the region that is close to the antenna. In this region, the electrical and magnetic fields are often complicated and are difficult to measure. This region is defined as

$$R < 0.62\sqrt{\frac{D^3}{\lambda}} \tag{1.11}$$

Between the far-field and reactive near-field region is the radiative near-field, which is also referred to as the Fresnel region. In this region there are no reactive field components from the antenna and the radiating fields begin to emerge. Figure 1.4 illustrates these regions of the antenna.

Figure 1.5 shows the far-field radiation patterns of some typical antennas. The directivity (D) is defined as the radiated power per unit solid angle compared to what would be received by an isotropic radiator [3]. As shown in Figure 1.5, different types of antenna have different directivity. The microstrip patch normally has a broad radiation pattern with moderate directivity. The dipole has an omnidirectional radiation pattern with typical gain of 2.3 dBi. The radiation pattern of the horn antenna has higher directivity. The directivity of the antenna can be increased by using an array antenna, as shown in Figure 1.5d.

The half-power beamwidth (HPBW) of an antenna is an important parameter for many applications, such as the base station antenna. It shows the angular range where

Figure 1.4 The different regions of the antenna field.

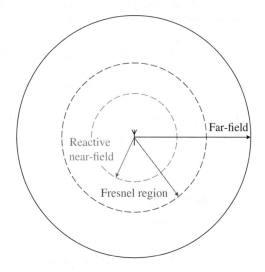

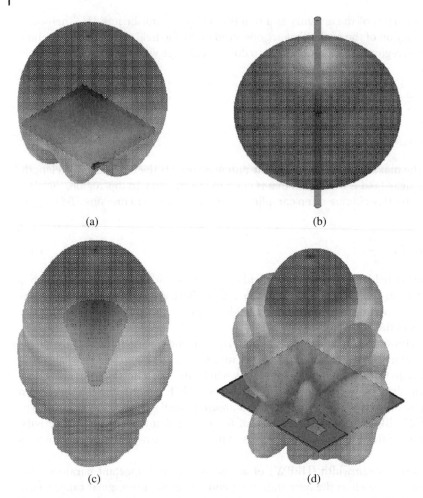

(a)

(b)

(c)

(d)

Figure 1.5 The radiation patterns of some typical antennas: (a) microstrip patch, (b) dipole antenna, (c) circular horn, and (d) microstrip array.

the radiated power has dropped by 50%. When plotting the radiation pattern in dB scale, the HPBW is where the power is reduced by 3 dB. Figure 1.6 shows an example of a directional radiation pattern that is plotted as the radiated power in dB versus the elevation angle (θ). High-directivity antennas always have a narrow HPBW, which can be seen from Figure 1.5.

The directivity of the antenna is calculated by

$$D(\phi, \varphi) = \frac{r^2 \frac{1}{2} Re[E \times H^*]}{P_{rad}/4\pi} \tag{1.12}$$

where P_{rad} is the radiated power. For a directional antenna, the directivity of the antenna can be estimated by [1]

$$D_0 \simeq \frac{4\pi(180/\pi)^2}{\theta_{1d}\theta_{2d}} = \frac{41253}{\theta_{1d}\theta_{2d}} \tag{1.13}$$

where θ_{1d} and θ_{2d} are the HPBW in two orthogonal planes.

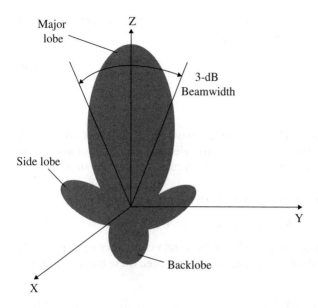

Figure 1.6 Illustration of the HPBW and the lobes of the antenna.

The gain of the antenna (G) is

$$G(\phi, \varphi) = \eta D(\phi, \varphi) \tag{1.14}$$

where η is the radiation efficiency of the antenna. The radiation efficiency describes how much input power is radiated from the antenna and is

$$\eta = \frac{P_{rad}}{P_{in}} \tag{1.15}$$

where P_{in} is the input power and P_{rad} is the radiated power. It is always desirable to have high-efficiency antennas; however, there are always some losses associated with the antenna, such as mismatching, dielectric loss, and conductor loss. The overall efficiency can be written as [1]

$$\eta = \eta_r \eta_c \eta_d \tag{1.16}$$

where η_r is the mismatching efficiency, η_c is the conduction efficiency, and η_d is the dielectric efficiency. The η_c and η_d are related to the material and are frequency dependent. The η_r can be calculated by

$$\eta_r = (1 - |\Gamma|^2) \tag{1.17}$$

When the mismatching efficiency is considered during the calculation, the calculated antenna gain is called the realised gain $(G_{realised})$. This is the overall efficiency of the antenna. Another method to define the antenna efficiency is to use the effective aperture, and the antenna efficiency is defined as the ratio of the effective area aperture to the actual physical size of the antenna

$$\eta = \frac{A_{eff}}{A} \tag{1.18}$$

where A represents the physical aperture size and A_{eff} represents the effective aperture size of the antenna. The A_{eff} can be calculated by

$$A_{eff} = \frac{\lambda^2}{4\pi} G \tag{1.19}$$

1.2.3 Polarisations

The polarisation of an antenna is defined as the polarisation of the wave radiated by the antenna [1]. Depending on the orientation of the electric field, the polarisation of the antenna can be classified as linearly polarised, circularly polarised or elliptically polarised.

Assume a plane wave travelling in the $-z$ direction, which can be written as

$$\vec{E}(z, t) = \vec{x} E_{x0} \cos(\omega t + kz + \phi_x) + \vec{y} E_{y0} \cos(\omega t + kz + \phi_y) \tag{1.20}$$

where E_{x0} and E_{y0} are the maximum magnitudes of the x and y components, respectively. The antenna is linearly polarised if the phase difference between these two components is $180°$ or

$$\Delta\phi = |\phi_x - \phi_y| = n\pi, n = 0, 1, 2, ... \tag{1.21}$$

If the amplitudes of the x and y components are the same while the phase difference $\Delta\phi$ is $90°$, the antenna is circularly polarised

$$\Delta\phi = |\phi_x - \phi_y| = n\pi/2, n = 1, 3, 5, ... \tag{1.22}$$

$$E_{x0} = E_{y0} \tag{1.23}$$

For a circularly polarised wave, the electric vector at a given point in space traced as a function of time is a circle. The sense of rotation can be determined by observing the direction of the field's rotation as the wave is viewed along the direction of propagation. If the rotation is clockwise, the wave is right-hand circularly polarised (RHCP). If the field rotation is anti-clockwise, the wave is left-hand circularly polarised (LHCP).

If the amplitudes of the x and y components are not equal but the phase difference $\Delta\phi$ is $90°$, then the antenna is elliptically polarised

$$\Delta\phi = |\phi_x - \phi_y| = n\pi/2, n = 1, 3, 5, ... \tag{1.24}$$

$$E_{x0} \neq E_{y0} \tag{1.25}$$

For the elliptical polarisation, the electric vector traced at a given position is a tilted ellipse, as shown in Figure 1.7. In practice, it is impossible to obtain a pure circularly polarised antenna within the entire bandwidth of the antenna. Thus, the term axial ratio (AR) is defined to describe how circular the radiated wave is. It is defined as the ratio of the major axis to the minor axis of the ellipse [1]

$$AR = \frac{OA}{OB} \tag{1.26}$$

where OA represents the major axis and OB represents the minor axis of the ellipse. These can be calculated by using the following equations

$$OA = \left[\frac{1}{2}(E_{x0}^2 + E_{y0}^2) + [E_{x0}^4 + E_{y0}^4 + 2E_{x0}^2 E_{y0}^2 \cos(2\Delta\phi)]^{1/2} \right]^{1/2} \tag{1.27}$$

$$OB = \left[\frac{1}{2}(E_{x0}^2 + E_{y0}^2) - [E_{x0}^4 + E_{y0}^4 + 2E_{x0}^2 E_{y0}^2 \cos(2\Delta\phi)]^{1/2} \right]^{1/2} \tag{1.28}$$

Figure 1.7 Tilted ellipse of elliptical polarisation.

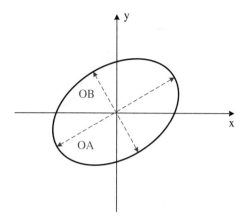

The tilt angle of the ellipse relative to the y axis is

$$\tau = \frac{\pi}{2} - \frac{1}{2}\tan^{-1}\left[\frac{2E_{x0}E_{y0}}{E_{x0}^2 - E_{y0}^2}\cos(\Delta\phi)\right] \tag{1.29}$$

AR is an important parameter of a circularly polarised antenna. Normally it is required that the AR of a circularly polarised antenna at the frequency band of interest is below 3 dB. In some applications, such as satellite communications, the requirement for the AR is more rigorous. For a circularly polarised antenna, it is important to check both the impedance and AR bandwidth, as they do not necessary overlap in the same frequency range. As an example, Figure 1.8a shows the impedance and AR bandwidth of a circularly polarised patch antenna. The AR is taken at the angle $\theta = 0°$ where the maximum gain of the patch is. This patch is a probe-fed square patch and has resonance at 7.9 GHz. The patch is corner truncated in order to obtain circular polarisation. Figure 1.8b shows the layout of this circularly polarised patch. The AR minimum is at 7.5 GHz and the AR ($AR < 3$) bandwidth partially overlaps with the impedance bandwidth of the patch.

Besides the bandwidth, another important parameter for a circularly polarised antenna is the AR beamwidth. The AR beamwidth describes the coverage region where the radiated waves from the antenna are circularly polarised. Figure 1.9 shows the simulated radiation pattern of the circularly polarised patch at its resonance in the $\phi = 0°$ plane. As shown, the dominant polarisation of this patch is RHCP and the 3 dB beamwidth of the patch is 80° (from −40° to +40°). The 3 dB AR beamwidth corresponds to the angle range where the LHCP is 15 dB lower than RHCP, which is from −31° to +85°. Thus, the overlapped HPBW and AR beamwidth is from −31° to +40°.

1.3 Antenna Array Fundamentals

An antenna array consists of multiple antenna elements positioned within an aperture. The total radiation pattern of an array antenna is the vector addition of the field radiated by each radiating element, and it can be expressed as the product of the array factor (AF) times the radiation pattern of the isolated radiating element. Each array element is excited by the input RF signal with certain amplitude and phase. By controlling the

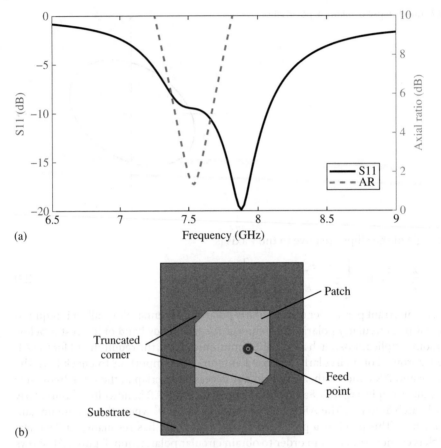

(a)

(b)

Figure 1.8 (a) The return loss and AR of a patch antenna. (b) The layout of the corner truncated square patch.

excitations of the array antenna, the beams of the array antenna can be formed to provide constructive interference to the desired direction and form nulls at undesired directions.

The array antenna is a critical component of the smart antenna. It determines the beam-steering performance (e.g. the largest beam scanning angle), directivity, and radiation efficiency of the antenna system. The BFN and the microwave circuit determine the beam-steering or beam-switching capability of the array antenna. The antenna array can be either a linear array or a planar array. The linear array is an array of antennas placed along one axis, as shown in Figure 1.10. A number of identical elements are spaced by a distance d and excited with phase difference β. Assuming that these elements are excited by identical amplitudes, the array factor can be derived as [1]

$$AF = \sum_{n=1}^{N} e^{j(n-1)\psi} \qquad (1.30)$$

where $\psi = kd \cos \theta + \beta$ and θ is the scan angle of the array antenna.

Besides the linear array configuration, antenna elements can also be placed along a rectangular grid to form a planar array, as shown in Figure 1.11. Planar arrays are more flexible in beamforming and, ideally, the main beam can be steered to any desired

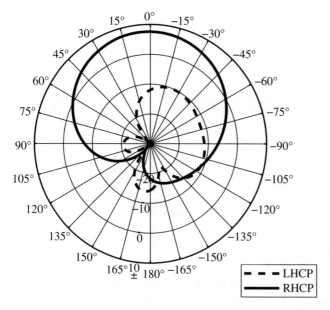

Figure 1.9 The simulated radiation pattern of the circularly polarised patch at its resonance in the $\phi = 0°$ plane.

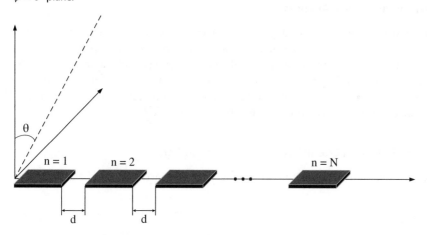

Figure 1.10 Linear array configuration.

direction (θ, ϕ). Thus, in general, the planar array is preferred for the design of smart antennas so it can benefit from the advanced BFN and signal processing algorithm. The array factor of the planar array can be written as [1]

$$AF = S_{xm}S_{yn} \qquad (1.31)$$

where

$$S_{xm} = \sum_{m=1}^{M} e^{j(m-1)(kd_x \sin \theta \cos \phi + \beta_x)} \qquad (1.32)$$

$$S_{yn} = \sum_{n=1}^{N} e^{j(n-1)(kd_y \sin \theta \sin \phi + \beta_y)} \qquad (1.33)$$

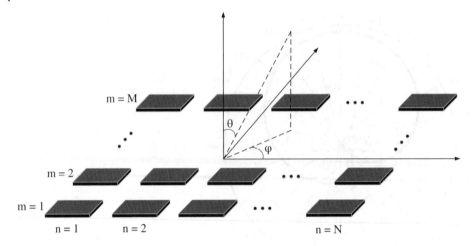

Figure 1.11 Planar array configuration.

d_x and d_y are the distances between the elements in the x and y axes, and β_x and β_y are the phase differences between the elements in the x and y axes, respectively.

1.3.1 Array Performance Analysis

The performance of the array is dependent on the radiation characteristics of the antenna element. There are also other factors that can affect the performance of the array antenna. For example, due to mutual coupling between adjacent elements, the input impedance and radiation pattern of the antenna elements are different from the case where the antenna radiates alone in free space. Therefore, it is a necessary step to analyse the array performance and optimise the array unit cell.

1.3.2 Active Reflection Coefficient and Mutual Coupling

When the array elements are excited, the voltages on each array element are expressed as [4]

$$V_m = \sum Z_{mn} I_n \tag{1.34}$$

where Z_{mn} is the impedance matrix of the whole array. When $m = n$, this represents the self impedance of element n and when $m \neq n$, this represents the mutual impedance between element m and n. I_n represents the amplitude of the current on the element n and V_m represents the applied voltage on the element m. If there is no coupling between the array element m and n, Z_{mn} equals zero and the mutual impedance matrix $[Z_{mn}]$ is diagonal. In general, the mutual coupling between array elements should be minimised in order to reduce the effect of the mutual impedance variations when the array scans [5].

To demonstrate the mutual coupling between the adjacent elements, Figure 1.12 shows the simulated E-field of a two-element array where only one of the antenna elements is excited while the other one is terminated by a matched load. As shown, the radiated field of the antenna is electromagnetically coupled to the other element, which

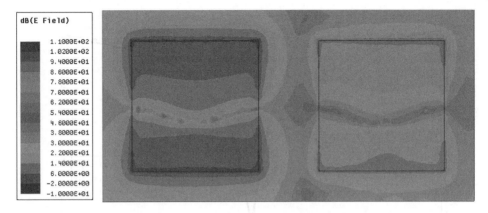

dB(E Field)

1.1000E+02
1.0200E+02
9.4000E+01
8.6000E+01
7.8000E+01
7.0000E+01
6.2000E+01
5.4000E+01
4.6000E+01
3.8000E+01
3.0000E+01
2.2000E+01
1.4000E+01
6.0000E+00
-2.0000E+00
-1.0000E+01

Figure 1.12 The mutual coupling in a two-element array antenna.

has an effect on the input impedance of the element. When there is a large number of elements presented, the mutual couplings are more complicated. One effective way to analyse and predict the radiation performance of the array element is to use the Floquet analysis [6].

In the Floquet analysis, superposition of Floquet modal functions is used to represent an array antenna of infinite size. This method is based on the fact that an infinite array antenna is equivalent to an array of infinite source functions. By performing Fourier transform and using the periodic function, Floquet series and Floquet sources are derived. Although the array antenna has a finite size, the analysis of an infinite array is still useful because the antenna elements in the central region of an electrically large array show similar radiation characteristics to the element in an infinite array after taking consideration of the mutual coupling effects. Moreover, it was shown in [6] that the performance of a finite array can be determined accurately by utilising the results derived in an infinite array. Floquet analysis can be performed in electromagnetic (EM) simulations by designing an array unit cell and assigning it periodic boundary conditions. This provides a good estimation of the input impedance matching, active reflection coefficient and scanning loss of the array antenna without using large computation resources. Figure 1.13 shows the simulated reflection coefficient of an array unit cell at different scan angles. The array unit cell is a square patch resonating at 30 GHz that is simulated using the Floqut port and periodic boundary conditions in a three-dimensional EM simulator. As shown in Figure 1.13, its reflection coefficient changes with different scan angles. This is due to the variation of the mutual couplings with different scan angle, which affects the input impedance of the array element.

Figure 1.14a shows the configuration of a 3×3 array using the square patch as the radiating element. Figure 1.14b compares the simulated active reflection coefficient of the central antenna element with the case when it radiates as an isolated element. The active reflection coefficient is obtained by activating all the nine antennas, while in the other case the reflection coefficient is obtained by only activating the central element and terminating the other elements by 50 Ω loads. As shown, the resonance of the patch shifts to a higher frequency when all the patches are active. Thus, some re-tunings, for example resizing the width of the patch, are required in order to shift the resonance back to the desired frequency.

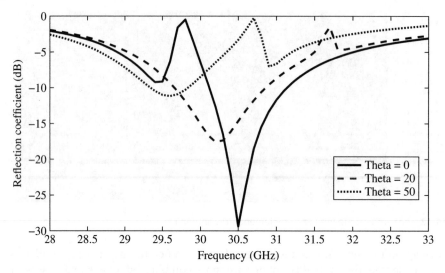

Figure 1.13 Simulated reflection coefficient of an array unit cell with different scan angles.

1.3.3 Directivity and Beamwidth

The directivity (D) of the array antenna is decided by the number of radiating elements when the distance between each array element is fixed [4]

$$D = \frac{4\pi A}{\lambda^2}\varepsilon_A \cos \theta \tag{1.35}$$

where A is the size of the radiating aperture and ε_A is the aperture efficiency of the array. If the beam has a symmetrical pattern in both the E- and H-planes, the 3 dB beamwidth of the array antenna in radians with uniform illumination taper can be estimated by

$$\theta_{3dB} = \sqrt{\frac{4\pi 0.866^2}{D}} \tag{1.36}$$

With the phase and amplitude errors, there would be pointing error, increased sidelobe level and directivity decrease. When phase and amplitude errors present, the directivity of an array antenna and the variance of the beam pointing deviation can be estimated by [4]

$$\frac{D}{D_0} = \frac{P}{1 + \overline{\sigma}^2 + \overline{\phi}^2} \tag{1.37}$$

$$\overline{\Delta}^2 = \overline{\phi}^2 \frac{\sum I_i^2 x_i^2}{\left(\sum I_i x_i^2\right)^2} \tag{1.38}$$

where D is the directivity of the array with errors, D_0 is the directivity of the array without any phase/amplitude errors, P is the probability that the error happens, ϕ is the phase error, σ is the amplitude error, $\overline{\Delta}$ is the variance of beam pointing deviation, I_i is the amplitude of the ith element excitation, and x_i is the element position divided by element spacing d. Due to the existence of these errors, a smart antenna with a BFN is required to perform calibrations before it is equipped with the communication system.

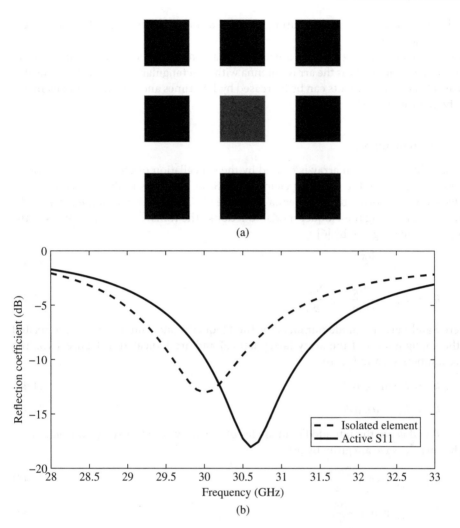

Figure 1.14 (a) Layout of a 3 × 3 array antenna. (b) Comparison between the active and isolated reflection coefficients of the array antenna element.

1.3.4 Grating Lobe

The additional main beam is observed when the spacing between the array elements is large because larger spacing allows the radiated waves from each element to add in phase at other angles besides the main beam of the array [7]. This is called the grating lobe.

The maximum scan angle of a two-dimensional array using a rectangular grid without any grating lobes is calculated by [6]

$$p = \frac{\lambda_0}{1 + \sin\theta_{max}} \tag{1.39}$$

where p is the period of the elements, λ_0 is the free space wavelength and θ_{max} is the maximum scan angle of the array antenna without any grating lobes. Generally speaking, in order to avoid the grating lobe, the distance between the elements is kept as a half-wavelength.

Given the same grating lobe free area, an array antenna with hexagonal lattice requires 0.866 as many elements as the array antenna with a rectangular lattice [7]. Thus, the distance between the elements can be increased by 1.15 times and the number of elements can be reduced by 15%.

1.3.5 Scan Blindness

The scan blindness of an array is caused by the cancellation of the dominant mode by the higher modes of the radiating element [7]. Scan blindness is also related to a true surface wave supported by the antenna structure [4]. The blind angle of the array can be determined by using the Floquet modal functions. The transverse wavenumbers of the Floquet mode are given by [6]

$$k_{xij} = k_{x0} + \frac{2i\pi}{a} \tag{1.40}$$

$$k_{yij} = k_{y0} - \frac{2i\pi}{a \tan \gamma} + \frac{2j\pi}{b} \tag{1.41}$$

where i and j are the mode numbers for the Floquet mode, and a, b and γ are related to the configuration of the array (array lattice) and are indicated in Figure 1.15. The wavenumbers k_{x0} and k_{y0} are

$$k_{x0} = k_0 \sin \theta \cos \phi \tag{1.42}$$

$$k_{y0} = k_0 \sin \theta \sin \phi \tag{1.43}$$

where θ and ϕ are the intended scan angles of the array. The transverse wavenumbers of the surface wave are given by [6]

$$k_{xs} = k_{ps} \cos \alpha + \frac{2m\pi}{a} \tag{1.44}$$

$$k_{ys} = k_{ps} \sin \alpha + \frac{2n\pi}{b} \tag{1.45}$$

where α is the propagation angle with the x axis of the surface wave. The k_{ps} is estimated by the following expression when the antenna elements are printed on a relatively thin substrate [8]

$$k_{ps} \approx k_0[1 + 0.5(k_0 h(1 - \frac{1}{\varepsilon_r}))^2] \tag{1.46}$$

Scan blindness occurs when the wavenumber of the surface wave matches the wavenumber of the Floquet mode. This means that there is a strong coupling between the surface wave and the Floquet mode. Since a rectangular grid is employed in this study and it is known that (0,0) is the dominant Floquet mode, the following equations are derived from [6]

$$k_{ps} \cos \alpha + \frac{2m\pi}{a} = k_0 \sin \theta_{blind} \cos \phi \tag{1.47}$$

$$k_{ps} \cos \alpha + \frac{2n\pi}{b} = k_0 \sin \theta_{blind} \sin \phi \tag{1.48}$$

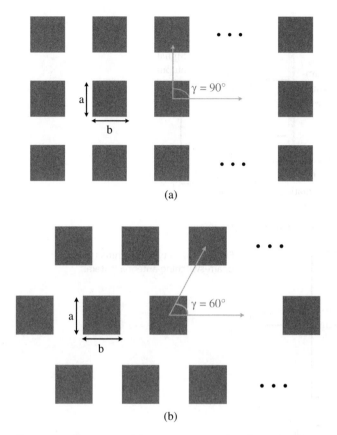

Figure 1.15 Illustration of the parameters used in Floquet mode analysis: (a) rectangular lattice and (b) hexagonal lattice.

where θ_{blind} is the scan blindness angle, and m, n is the mode number. At the scan blindness angle, all of the power incident on the array is trapped in the non-radiating surface wave, which results in total reflection ($|S11| \approx 0\ dB$) [9]. As shown in Figure 1.13, total reflection takes place at certain frequencies when the array scans, which means that scan blindness would occur at these frequencies and scan angles.

1.4 Smart Antenna Architecture and Hardware Implementation

A smart antenna consists of an array antenna and a signal processing unit. The signal processing unit uses some advanced algorithms to identify the DOA of the signal and calculate beamforming vectors, which steer the beam of the array antenna to the desired direction and suppress the lobes at unwanted angles. Figure 1.16 shows the generic system configuration of a smart antenna.

Smart antennas can be categorised into two main types: beam-switching antenna systems and adaptive array antenna systems. The beam-switching smart antenna has a BFN

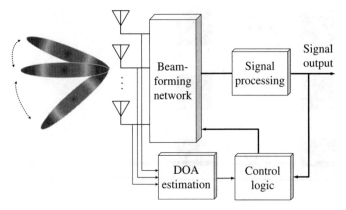

Figure 1.16 The generic system configuration of a smart antenna.

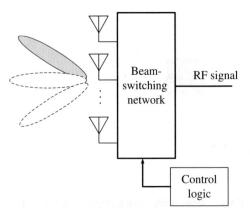

Figure 1.17 A typical architecture for a beam-switching antenna system.

that provides the fixed phase shift to form shaped beams in pre-defined directions. Figure 1.17 shows a typical architecture for a beam-switching smart antenna. It consists of an antenna array, a beamforming feed network and a control unit for beam selection. The antenna array can be either a linear or a planar array, depending on the requirements of the targeted applications. The beamforming feed network is a critical component because it determines the number of available beams that can be switched. The most common techniques for designing a beam-switching network include the use of the Butler matrix, introducing microwave switches such as PIN diodes or RF micro-electromechanical system (MEMS) switches on the feed network [10–12]. Since they do not need to employ any phase shifters and their control circuits are relatively simple, beam-switching smart antennas provide a low-cost and low-power consumption solution if continuous beam-steering is not required. The main disadvantage for the beam-switching smart antenna is that it can only provide a limited number of beams. For example, the parasitic array antenna only has one driven element surrounded by several parasitic elements. Beam-switching is obtained by controlling the reactive loads of the parasitic elements. The number of achievable beams is limited and is determined by the number of parasitic elements as well as the adjustable range of the reactive loads.

Adaptive smart antennas are able to continuously steer the beam to the desired directions and shape the beam to maximise the link budget with sophisticated

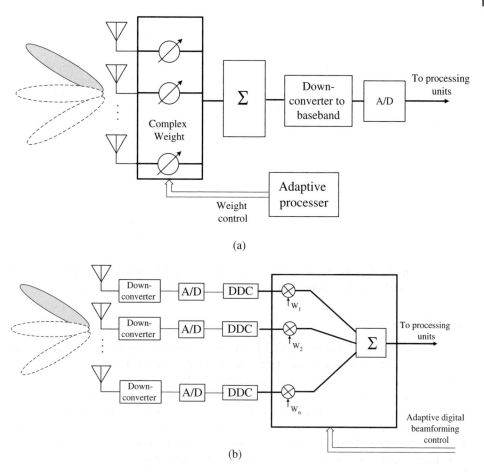

Figure 1.18 The generic system architecture of: (a) the smart array antenna using analogue beamforming and (b) digital beamforming.

signal processing algorithms. The beamforming of the adaptive array antennas can be achieved by two approaches: analogue or digital. Figure 1.18 shows the generic system architecture of the smart array using analogue beamforming (ABF) and digital beamforming (DBF).

ABF requires the use of a large number of phase shifters and attenuators to provide complex weights to the array elements. The summed signals need to be down-converted to baseband signals then pass to the analogue/digital (A/D) converter before being sent to the signal processing unit. Instead of using a large number of microwave phase shifters, which are expensive and lossy, DBF uses an A/D converter and a digital down-converter (DDC) on each array element and then the beamforming can be realised digitally. However, DBF requires a large number of A/D converters, which have high power consumption and need real-time signal processing. To realise adaptive beamforming, there are several well-defined algorithms available, such as the LMS algorithm, the recursive least squares algorithm and the CMA. Some low-cost beamforming algorithms are summarised in Chapter 2 and readers can refer to this

Table 1.1 Comparison of the beam-switching and adaptive arrays.

	Beam-switching array	Adaptive array
Cover range	Larger than traditional antenna system	Larger than beam-switching antenna system
Signal processing	Basic	Advanced
System complexity	Low	High
Beamforming	Limited	Advanced
Power consumption	Low/medium	High
Interference rejection capability	Poor	Good
Cost	Low	High

chapter for more details. Table 1.1 compares the beam-switching array and adaptive array in terms of the performance, system complexity and cost.

1.4.1 ADC and DAC

The analogue-to-digital converter (ADC) converts an analogue signal into a digital signal while the digital-to-analogue converter (DAC) converts a digital signal into an analogue signal. The ADC is used in the receiving system and the DAC is used at the transmitter. Resolution is one of the key parameters for the ADC, as it determines the quantisation error. For example, an ADC with a resolution of 6 bits can encode an analogue input to one in 64 different levels ($2^6 = 64$). Ideally, the quantisation error for the ADC should be as small as possible. However, a higher resolution converter normally has a higher cost.

Another important parameter of the ADC is the sampling rate. Based on Nyquist sampling theorem, a signal should be sampled at a rate greater than twice its maximum frequency component. For a wideband antenna system, this means that it needs a very high sampling rate. A high sampling rate ADC needs to consume more power and the cost is high. For a low-cost system, the undersampling technique can be used. The advantages of undersampling include low power consumption, low cost, easy capturing of ADC data, and easy interface to field-programmable gate arrays (FPGAs) (because of its lower speeds) [13].

1.4.2 Digital Down-Converter (DDC)

The DDC converts the digitised signal (e.g. the output of the ADC) to a lower frequency signal at a lower sampling rate. The DDC is normally used in the digital beamforming system, where the received signal from each of the antenna elements is down-converted and converted to the digital signals.

1.4.3 Digital Signal Processor

A digital signal processor (DSP) is a microprocessor that processes the digitalised signals and performs the mathematical manipulation. The DSP is important hardware for the smart antenna because it performs signal processing and beamforming. A DSP contains the program memory, data memory, compute engine, and input/output [14]. The program memory stores the programs that can be used to perform the signal processing, such as the beamforming algorithms, while the data memory saves the information to be processed. The compute engine processes the digital signal using the program saved in the program memory and the data stored in the data memory. For a smart antenna with a sophisticated signal processing algorithm, the DSP needs to have large computation resource and consume more power. A typical example is the digital BFN.

1.4.4 Field-programmable Gate Array

The development of the FPGA provides an alternative way to design the BFN of smart antennas [15, 16]. FPGAs contain an array of programmable logic blocks and can be reconfigured for different applications. They have embedded processors, which provide options for implementing various adaptive signal processing algorithms such as DOA estimation, MUltiple SIgnal Classification (MUSIC), and the LMS algorithm. It is also possible to have a PFGA board with built-in A/D converter and other modules as well. Figure 1.19 is a diagram of a smart antenna using FPGA.

To reduce the cost, the price of a high-end PFGA and related modules puts pressure on the development of low-cost smart antennas. Using some external modules instead of embedding the modules (e.g. external ADC) on the FPGA reduces the cost of the system. A low-cost hardware implementation of a smart antenna based on FPGA was reported in [17]. The beamforming algorithm was implemented on a low-end FPGA and the adaptive beamforming was realised by using the CMA.

Figure 1.19 A block diagram of a smart antenna using FPGA.

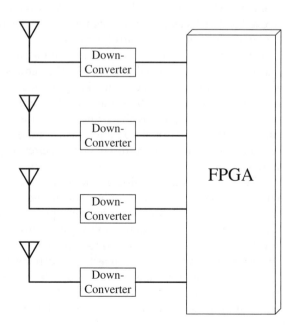

1.5 Overview of the Book

The book contains seven chapters. Chapter 1 presents a brief discussion on antenna and array antenna fundamentals, as well as smart antenna configurations. The aim of the chapter is to provide readers with some basic concepts and fundamental theories of the RF part of smart antennas.

Chapter 2 discusses the basic principles of beamforming and introduces some representative beamforming methods and algorithms for smart antennas. A review of the particular area of low-cost adaptive beamforming is also presented in this chapter, including hybrid beamforming and robust adaptive beamforming. The objective of the chapter is to provide readers with some basics of beamforming algorithms and their applications in wireless commutations. A list of references is provided in the end of this chapter and interested readers can refer to these references for more details.

Chapter 3 presents low-cost smart antenna design using the electronically steerable parasitic array radiator (ESPAR). ESPAR is an attractive solution to the design of low-cost smart antenna because it has only one RF transceiver and the electronic beam scanning of the antenna system can be achieved without using any phase shifters. This chapter introduces various designs of the ESPAR using different types of antenna elements, such as monopole, microstrip slot antenna, and patch. Varactors or PIN diodes can be employed to load the parasitic elements of ESPAR, and many examples are discussed in this chapter. The link quality test is performed to validate channel improvement using the ESPAR.

Chapter 4 presents the design of beam-switching and beam-steering antenna using the active frequency selective surface (FSS). The FSS is a periodic structure that shows space filtering characteristics, for example bandpass or bandstop. By incorporating RF PIN diodes or varactors in the FSS, the characteristics of the FSS can be controlled. In this chapter, the fundamental theories of the FSS are presented first, then different approaches to realise a low-cost beam-switching and beam-steering antenna using active FSS are discussed with design examples, including some practical techniques to design the bias circuit to control the FSS.

In Chapter 5 the design of beam-scanning and beam-switching reflectarrays and transmitarrays is discussed. Both types of antennas are space-fed and they can operate without the use of the feed network. By placing the feeds in the focal arc, multibeam or beam switching can be obtained. They are suitable for the applications where beam switching, high directivity, and low cost are required. This chapter presents the fundamental theories and operation principles of the reflectarrays and transmitarrays with design examples. Some discussions on circularly polarised design are also included in this chapter.

MIMO antenna uses multiple antennas at the transmitter and receiver to increase channel throughput by taking advantage of the space diversity. Chapter 6 presents the MIMO theories and discusses different approaches to design compact low-cost MIMO antennas, such as the use of neutralising lines, decoupling networks, and metamaterials. A case study is given to demonstrate how to improve the isolation between two closely spaced planar inverted-F antennas for smartphones.

Chapter 7 presents some other types of low-cost smart antennas, including dielectric lenses, retrodirective arrays, Fabry–Perot resonant antennas, array-fed reflectors and multibeam antennas based on BFNs. The basic principles and design techniques of various types of low-cost smart antennas are explained, and many examples of recent developments in each type of smart antennas are discussed. These are useful for readers who need to understand different techniques in low-cost smart antennas in order to choose a specific type of smart antenna for relevant applications.

References

1 C.A. Balanis. *Antenna Theory: Analysis and Design*. John Wiley & Sons Inc., 3rd edition, 2015.

2 W.A. Imbriale, S. Gao, and L. Boccia, editors. *Space Antenna Handbook*. John Wiley & Sons Ltd, 2012.

3 J.D. Kraus and R.J. Marhelfa. *Antenna for all Applications*. McGraw-Hill, 2002.

4 R.J. Mailloux. *Phased Array Antenna Handbook*. Artech House Inc., 2005.

5 P.W. Hannan, D. Lerner, and G. Knittel. Impedance matching a phased-array antenna over wide scan angles by connecting circuits. *IEEE Transactions on Antennas and Propagation*, 13(1):28–34, Jan 1965.

6 A.K. Bhattacharyya. *Phased Array Antennas: Floquet analysis, synthesis, BFNs, and active array systems*. John Wiley & Sons Inc., 2006.

7 R.C. Hansen. *Phased Array Antennas*. John Wiley & Sons Inc., 2009.

8 W.C. Chew, J.A. Kong, and L.C. Shen. Radiation characteristics of a circular microstrip antenna. *Journal of Applied Physics*, 51(7):3907–3915, 1980.

9 D. Pozar and D. Schaubert. Scan blindness in infinite phased arrays of printed dipoles. *IEEE Transactions on Antennas and Propagation*, 32(6):602–610, Jun 1984.

10 W.F. Moulder, W. Khalil, and J.L. Volakis. 60-GHz two-dimensionally scanning array employing wideband planar switched beam network. *IEEE Antennas and Wireless Propagation Letters*, 9:818–821, 2010.

11 S.A. Mitilineos and C.N. Capsalis. A new, low-cost, switched beam and fully adaptive antenna array for 2.4 GHz ISM applications. *IEEE Transactions on Antennas and Propagation*, 55(9):2502–2508, Sept 2007.

12 C.E. Patterson, W.T. Khan, G.E. Ponchak, G.S. May, and J. Papapolymerou. A 60-GHz active receiving switched-beam antenna array with integrated butler matrix and GaAs amplifiers. *IEEE Transactions on Microwave Theory and Techniques*, 60(11):3599–3607, Nov 2012.

13 Texas Instruments. Why oversample when undersampling can do the job? Technical report, 2013.

14 Analog Devices. A beginner's guide to digital signal processing (DSP). Technical report, accessed in March 2018.

15 M. Kim, K. Ichige, and H. Arai. Real-time smart antenna system incorporating FPGA-based fast DOA estimator. In *IEEE 60th Vehicular Technology Conference, 2004. VTC2004-Fall. 2004*, volume 1, pages 160–164 Vol. 1, Sept 2004.

16 H. Arai and K. Ichige. *Hardware Implementation of Smart Antenna Systems for High Speed Wireless Communication*. Proceedings of 2005 URSI General Assembly, No. BC-6, New Delhi, 2005.

17 A. Chinatto, C. Junqueira, and J.M.T. Romano. Low cost smart antenna array hardware implementation. In *2011 SBMO/IEEE MTT-S International Microwave and Optoelectronics Conference (IMOC 2011)*, pages 784–788, Oct 2011.

2

Beamforming Algorithms for Smart Antennas

2.1 Introduction

The smart antenna technique, in most cases, is an application of the traditional digital beamforming technique in array signal processing to the area of wireless communications [1–9]. In this chapter we focus on some basic principles of beamforming, particularly digital beamforming, and introduce some representative beamforming methods and algorithms. First, we give a brief introduction to beamforming and then introduce some basic concepts in beamforming, such as steering vectors, spatial aliasing, beam response, and beam pattern. We then present two fixed beamfomer design methods, finite impulse response (FIR) filter based design and least squares based design, followed by two traditional adaptive beamforming algorithms: the reference signal based beamformer and the Capon beamformer. Next, we consider two specific applications with their associated blind adaptive beamforming algorithms: one for satellite navigation systems and one for wireless communications. Finally, we provide a review of the particular area of low-cost adaptive beamforming with two different approaches: hybrid beamforming, which employs analogue and digital beamformers at different stages, and robust adaptive beamforming, which focuses on two representative robust algorithms.

The aim of beamforming is to effectively estimate the signal of interest in the presence of noise and interferences employing an array of antennas, which are located at different spatial positions according to some specified geometry. Depending on their spatial layout, we can have linear arrays, planar arrays, volumetric arrays, etc. According to the regularity of their spacing, we can have either uniform arrays or non-uniform and sparse arrays. In particular, the study of various sparse array structures has become a hot topic due to their superior ability to estimate a very large number of signals using a small number of antennas [10–19]. Based on the construction of the individual antennas, we can have either traditional antenna arrays, where there is one single signal at the output of each antenna, or vector antenna arrays, where each whole individual 'antenna' is composed of multiple component antennas and there will be the same number of output signals as component antennas for each whole antenna. In the latter case, it looks like each whole antenna output is a vector of signals, hence the name vector antenna arrays or, for the general case, vector sensor arrays. Examples of vector antenna arrays include crossed dipole arrays and tripole arrays [20–28].

Low-cost Smart Antennas, First Edition. Qi Luo, Steven (Shichang) Gao, Wei Liu, and Chao Gu.
© 2019 John Wiley & Sons Ltd. Published 2019 by John Wiley & Sons Ltd.

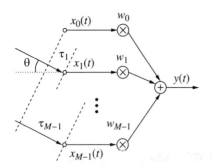

Figure 2.1 A beamforming structure based on a linear antenna array.

In this chapter we will only study the beamforming problem based on traditional linear array structures, but the ideas and principles introduced here can be easily extended to all other array structures mentioned above.

Beamforming is achieved by changing both the phase and the amplitude of the received antenna signals, which can be represented by applying complex-valued coefficients to the received signals, as shown in Figure 2.1, where M antennas sample the electromagnetic wave field spatially and the output $y(t)$ at time t is given by an instantaneous linear combination of these spatial samples $x_m(t)$, $m = 0, \ldots, M - 1$, as

$$y(t) = \sum_{m=0}^{M-1} x_m(t) w_m^* \tag{2.1}$$

where * denotes the complex conjugate and w_m, $m = 0, \ldots, M - 1$ are the beamformer coefficients. The complex conjugate operation in the expression is purely for the convenience of notations when calculating the covariance matrix and other second-order statistics of the signals. As a result of this notation, although w_m is called the beamformer coefficient, the effective one is w_m^*.

Note that although the received antenna signals are real by themselves, they can be treated as complex-valued signals after down-conversion to the baseband using a quadrature receiver, where the in-phase (I) and quadrature (Q) components of the receiver output are the real and imaginary parts of the complex-valued signals, respectively [29].

Moreover, for simplicity, we normally assume the array antennas have the same characteristics and they are omnidirectional (or isotropic), i.e. their response to an impinging signal is independent of its angles of arrival.

By designing the weight coefficients according to some criterion, such as maximising the output signal to interference plus noise ratio (SINR) or a mainlobe to sidelobe ratio, an overall specific spatial response is then achieved with 'beams' pointing to the desired signals and 'nulls' towards the interfering ones.

It should be noted that the beamforming structure shown in Figure 2.1 is mainly for narrowband signals and all the methods and algorithms discussed in this chapter are primarily for narrowband signals. For wideband signals we have to employ different structures, such as tapped delay-line or sensor delay-line based structures. Interested readers can refer to reference [9] for details of wideband beamforming.

In the next section we introduce some basic concepts related to beamforming.

2.2   Basic Concepts for Beamforming

The beam pattern of a beamformer describes its response to an impinging signal at a given frequency ω and angle of arrival θ. Consider a complex plane wave signal $e^{j\omega t}$ arriving from an angle θ, where $\theta \in [-\frac{\pi}{2} \ \frac{\pi}{2}]$ is measured with respect to the broadside of the linear array, as shown in Figure 2.1.

Assume the signal received by the first sensor is $x_0(t) = e^{j\omega t}$, and then by the mth sensor it is $x_m(t) = e^{j\omega(t-\tau_m)}$, $m = 1, \ldots, M-1$, where τ_m is the propagation delay for the signal from sensor 0 to sensor m, which is a function of θ. Then, the output $y(t)$ is given by

$$y(t) = e^{j\omega t} \left(\sum_{m=0}^{M-1} e^{-\omega \tau_m} w_m^* \right) \tag{2.2}$$

with $\tau_0 = 0$.

We can see that $y(t)$ is a scaled version of the input signal $e^{j\omega t}$ and the scalar is then the response of this beamformer, given by

$$P(\omega, \theta) = \sum_{m=0}^{M-1} e^{-j\omega \tau_m} w_m^* = \mathbf{w}^H \mathbf{d}(\omega, \theta) \tag{2.3}$$

where the weight vector $\mathbf{w}$ holds the M coefficients of the beamformer

$$\mathbf{w} = [w_0 \ w_1 \ \ldots \ w_{M-1}]^T \tag{2.4}$$

and the vector $\mathbf{d}(\omega, \theta)$ is given by

$$\mathbf{d}(\omega, \theta) = [1 \ e^{-j\omega \tau_1} \ \ldots \ e^{-j\omega \tau_{M-1}}]^T \tag{2.5}$$

$\mathbf{d}(\theta, \omega)$ is normally referred to as the array response vector, the steering vector or the direction vector of the array [30] and we will use the term 'steering vector' in our context.

In our notation, we generally use lowercase bold letters for vector valued quantities, while uppercase bold letters denote matrices. The operators $\{\cdot\}^T$ and $\{\cdot\}^H$ represent transpose and Hermitian transpose, respectively.

Next, we give a brief discussion of the spatial aliasing problem encountered in array processing.

In array processing, the sensors sample the impinging signals spatially and if the signals from different spatial locations are not sampled by the array sensors densely enough, i.e. the inter-element spacing of the array antennas is too large, then source signals from different directions will have the same array steering vector and we cannot uniquely determine their directions based on the received array signals. As a result, we have a spatial aliasing problem due to ambiguity in the DOAs of source signals.

For signals having the same angular frequency ω and the corresponding wavelength λ, but different DOAs θ_1 and θ_2 satisfying the condition $(\theta_1, \theta_2) \in [-\pi/2 \ \pi/2]$, aliasing implies that we have $\mathbf{d}(\theta_1, \omega) = \mathbf{d}(\theta_2, \omega)$, i.e.

$$e^{-j\omega \tau_m(\theta_1)} = e^{-j\omega \tau_m(\theta_2)} \tag{2.6}$$

For a uniformly spaced linear antenna array with an inter-element spacing d, we have $\tau_m = m\tau_1 = m\frac{d \sin \theta}{c}$ and $\omega \tau_m = m\frac{2\pi d \sin \theta}{\lambda}$. Then, (2.6) changes to

$$e^{-jm\frac{2\pi d \sin \theta_1}{\lambda}} = e^{-jm\frac{2\pi d \sin \theta_2}{\lambda}} \tag{2.7}$$

To avoid aliasing, the condition $|2\pi(\sin\theta)d/\lambda|_{\theta=\theta_1,\theta_2} < \pi$ has to be satisfied. Then, we have $|d/\lambda \sin\theta| < 1/2$. Since $|\sin\theta| \le 1$, this requires that the array distance d should be less than $\lambda/2$.

For the case of $d = \lambda/2$, we have $\omega\tau_m = m\pi\sin\theta$ and the beamformer response is given by

$$P(\omega,\theta) = \sum_{m=0}^{M-1} e^{-jm\pi\sin\theta} w_m^* \qquad (2.8)$$

which becomes a function of θ only.

For a given frequency ω and a set of values of w_m, $m = 0, \ldots, M-1$, we can use (2.3) to calculate the resultant amplitude response $|P(\theta,\omega)|$ and $|P(\theta,\omega)|$ is called the beam pattern of the beamformer to describe its sensitivity with respect to signals arriving from different directions and with different frequencies.

Since for most of the cases we are interested in, the impinging signals are narrowband with a fixed centre frequency, we can omit ω in $|P(\theta,\omega)|$ and simplify it into $|P(\theta)|$. For uniform linear arrays (ULAs) with half-wavelength spacing, we can use (2.8) to calculate the beam pattern directly.

As an example, consider a ULA with 10 antennas and uniform weighting, i.e. $w_m = 1$, $m = 0, \ldots, 9$. Its beam response is given by

$$P(\theta) = \sum_{m=0}^{9} e^{-jm\pi\sin\theta} \qquad (2.9)$$

Its beam pattern in dB is shown in Figure 2.2 and is defined as follows

$$BP = 20\log_{10}\frac{|P(\theta)|}{\max|P(\theta)|} \qquad (2.10)$$

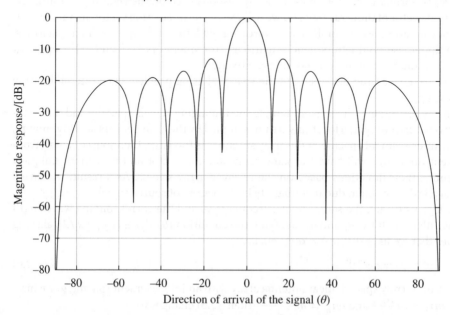

Figure 2.2 The beam pattern of the 10-antenna ULA-based beamformer with uniform weighting and half-wavelength spacing.

From Figure 2.2 we can see that it has a main beam (maximum magnitude response) pointing to the broadside ($\theta = 0°$), as all signals coming from the broadside are added coherently by the uniform weighting, while signals from other directions are not.

If we increase the spacing of the above beamformer from $\lambda/2$ to λ, but with the same uniform weighting, its beam response becomes

$$P(\theta) = \sum_{m=0}^{9} e^{-j2m\pi \sin\theta} \tag{2.11}$$

In Figure 2.3 we can see the aliasing problem very clearly, where in addition to the broadside main beam we also have the main beam pointing to the end-fire directions ($\pm 90°$).

The main beamwidth of the beamformer is dependent on the number of antennas and with more antennas, the beam becomes narrower. For the above example, we can increase the antenna number to 20 and again with uniform weighting its beam pattern is shown in Figure 2.4, where we can see that the broadside main beam is much sharper than the case with 10 antennas.

In this example, the beamformer coefficients have a fixed value and the response of the beamformer is fixed for a given direction, independent of the change in the operating environments. We call this kind of beamformer a *fixed beamformer* or a *data independent beamformer* [30]. On the other hand, we may need to adjust the beamformer's coefficients according to the received array signals to maximise the output SINR or some other criteria. This latter kind of beamformer is called an *adaptive beamformer* or a *data dependent beamformer* [30].

We will discuss some representative design methods for both fixed beamformers and adaptive beamformers in the following sections.

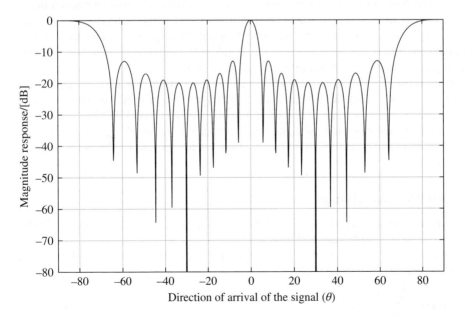

Figure 2.3 The beam pattern of the 10-antenna ULA-based beamformer with uniform weighting and one wavelength spacing.

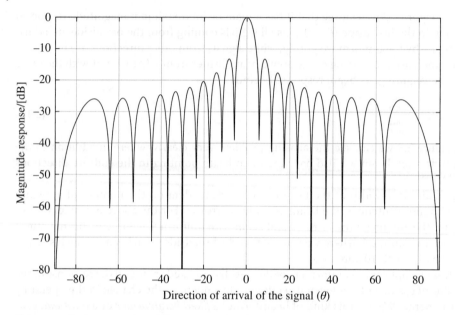

Figure 2.4 The beam pattern of the 20-antenna ULA-based beamformer with uniform weighting and half-wavelength spacing.

2.3 Fixed Beamformer Design

In this section, we study the design of fixed beamformers, which is also called array pattern synthesis. For fixed beamformer design we are normally given a desired beam pattern and the aim is to find the set of beamformer coefficients giving the best approximation to the desired pattern. This is a general optimisation problem and can be solved by employing various optimisation methods, such as the weighted Chebyshev approximation [31–33] and the convex optimisation methods [9, 34–39].

Here we introduce two simple but effective design methods. The first one deals with the special case of uniform linear arrays and exploits the similarity between the response of a ULA-based beamformer and that of a finite impulse response (FIR) filer. The second one is the least squares based design, which can deal with the design problem for a general antenna array. At the end of this section, we briefly discuss the beam steering problem.

2.3.1 FIR Filter Based Design

For a uniform linear array, its beam response is given by (2.8). The form is very similar to the response of a traditional FIR filter.

Recall that for an FIR filter with the same set of coefficients $w_m = 0$, $m = 1, \dots, 9$, its frequency response is given by [40]

$$P(\Omega) = \sum_{m=0}^{M-1} e^{-jm\Omega} w_m^* \tag{2.12}$$

with $\Omega \in [-\pi \ \pi]$ being the normalised frequency.

For the response given in (2.8), when θ changes from $-90°$ to $+90°$, $\pi \sin \theta$ changes from $-\pi$ to π accordingly, which is in the same range as Ω in (2.12). As a result, given a set of coefficients, we can calculate $P(\Omega)$ in (2.12) first, and then replace Ω by $\pi \sin \theta$ to obtain the beam response $P(\theta, \omega)$. Therefore, the design of ULA-based beamformers can be achieved by transforming the problem into an FIR filter design problem first and then applying existing FIR filter design approaches directly.

As an example, suppose we want to design a beamformer with its main beam pointing to directions $\theta \in [-\frac{\pi}{9} \; \frac{\pi}{9}]$ ($[-20° \; 20°]$), while suppressing signals from the remaining directions. Since $\sin \frac{\pi}{9} = 0.342$, it is equivalent to designing a lowpass filter with a bandpass $\Omega \in [-0.342\pi \; 0.342\pi]$. We use the window method to design such a filter and the MATLAB function *fir1* can be used for this design [41].

The following is the design result with $M = 12$

$$\mathbf{w} = [-0.0017, -0.0107, -0.0184, 0.0338, 0.1780, 0.3190,$$
$$0.3190, 0.1780, 0.0338, -0.0184, -0.0107, -0.0017]^T \qquad (2.13)$$

The resultant beam pattern is shown in Figure 2.5, where we can see that its main beam has pointed to the desired range of directions.

For the more general case of $d = \alpha\lambda/2$, $\alpha \leq 1$ is a scalar, and the response of the ULA-based beamformer given in (2.8) becomes

$$P(\theta) = \sum_{m=0}^{M-1} e^{-jm\alpha\pi \sin \theta} w_m^* \qquad (2.14)$$

Its design can be obtained in a similar way and the only difference is that the FIR filter can have an arbitrary response over the regions $\Omega \in [-\pi - \alpha\pi]$ and $[\alpha\pi \; \pi]$ without

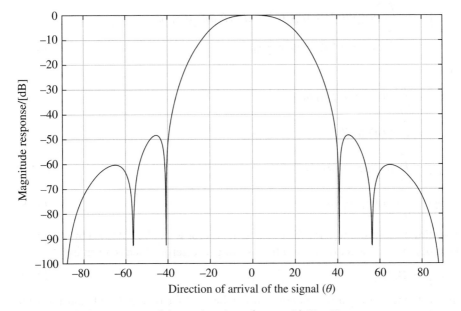

Figure 2.5 The beam pattern of the resultant beamformer with $M = 12$ antennas.

affecting that of the beamformer. However, it will affect the white noise gain of the beamformer [35–37], and more advanced techniques are needed to address this problem [9].

2.3.2 Least Squares Based Design

The FIR filter based approach is only applicable to ULAs. For a general array, a simple solution is the classic least squares based approach, which can give a closed-form solution to the beamformer coefficients.

The general least squares problem has a long history and has been studied very well, with numerous variations and solutions proposed [9, 42–49]. Given the desired beam response $P_d(\theta)$, the least squares based design is formulated by minimising the sum of the squares of the error between $P_d(\theta)$ and the designed response $P(\theta)$ over the whole interested range Θ of DOA angle θ

$$\min_{\mathbf{w}} \int_{\Theta} |P(\theta) - P_d(\theta)|^2 d\theta \tag{2.15}$$

We can add a pre-defined weighting function $v(\theta)$ to give different weighting for errors in different directions and form a weighted least squares problem

$$\min_{\mathbf{w}} \int_{\Theta} v(\theta)|P(\theta) - P_d(\theta)|^2 d\theta \tag{2.16}$$

The cost function in the above can be expanded into the following form

$$
\begin{aligned}
J_{ls}(\mathbf{w}) &= \int_{\Theta} v(\theta)|P(\theta) - P_d(\theta)|^2 d\theta \\
&= \int_{\Theta} v(\theta)(P(\theta) - P_d(\theta))(P(\theta) - P_d(\theta))^H d\theta \\
&= \int_{\Theta} v(\theta)(|P(\theta)|^2 + |P_d(\theta)|^2 - 2Re\{P(\theta)P_d^*(\theta)\}) d\theta \\
&= \mathbf{w}^H \mathbf{G}_{ls} \mathbf{w} - \mathbf{w}^H \mathbf{g}_{ls} - \mathbf{g}_{ls}^H \mathbf{w} + g_{ls}
\end{aligned} \tag{2.17}
$$

where

$$
\begin{aligned}
\mathbf{G}_{ls} &= \int_{\Theta} v(\theta)(\mathbf{d}(\theta)\mathbf{d}^H(\theta)) d\theta \\
&= \int_{\Theta} v(\theta)\mathbf{D}(\theta) d\theta \\
\mathbf{g}_{ls} &= \int_{\Theta} v(\theta)(\mathbf{d}(\theta)P_d^*(\theta)) d\theta \\
g_{ls} &= \int_{\Theta} v(\theta)|P_d(\theta)|^2 d\theta
\end{aligned} \tag{2.18}
$$

The optimum solution $\mathbf{w}_{opt}$ for the coefficients vector $\mathbf{w}$ can be obtained by taking the gradient of J_{ls} with respect to $\mathbf{w}^H$ and then setting it to zero, which is given by

$$\mathbf{w}_{opt} = \mathbf{G}_{ls}^{-1} \mathbf{g}_{ls} \tag{2.19}$$

Suppose the weighting function $v(\theta)$ is α for the sidelobe region $\theta \in \Theta_s$ and $1 - \alpha$ for the mainlobe region $\theta \in \Theta_m$, and the desired response $P_d(\theta)$ is one for the mainlobe region and zero for the sidelobe region. We also approximate the integration operations in (2.17) by discrete summations. Then, the optimum solution becomes

$$\mathbf{w}_{opt} \approx \mathbf{G}_D^{-1} \mathbf{g}_D \tag{2.20}$$

where

$$\mathbf{G}_d = (1 - \alpha) \sum_{\theta_k \in \Theta_m} \mathbf{D}(\theta_k) + \alpha \sum_{\theta_k \in \Theta_s} \mathbf{D}(\theta_k)$$

$$\mathbf{g}_D = (1 - \alpha) \sum_{\theta_k \in \Theta_m} \mathbf{d}(\theta_k) \tag{2.21}$$

As an example, we consider the following design problem for a ULA with $M = 12$ antennas and $\alpha = 0.5$. The mainlobe direction is the broadside, i.e. $\theta_0 = 0°$, and Θ_{ml} only includes one single direction θ_0. The sidelobe area is from $-90°$ to $-20°$ and $20°$ to $90°$, and is discretised into 100 points, 50 points for each side.

The design result is shown in Figure 2.6 and the coefficient vector is given by

$$\mathbf{w} = [0.0221, 0.0648, 0.1278, 0.2030, 0.2707, 0.3113,$$
$$0.3113, 0.2707, 0.2030, 0.1278, 0.0648, 0.0221]^T \tag{2.22}$$

2.3.3 Beam Steering

In the two fixed beamformer design examples provided earlier, the main beam is set to the broadside. However, the two introduced methods can be employed to design the main beam pointing to arbitrary directions, not limited to the broadside case.

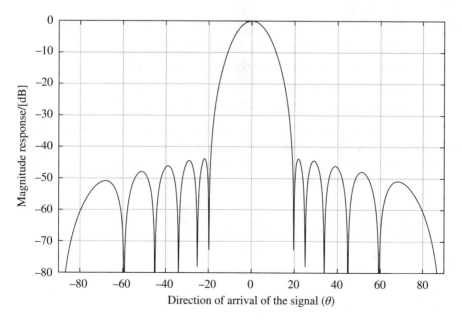

Figure 2.6 The resultant beam response for the design example using the least squares approach in (2.20).

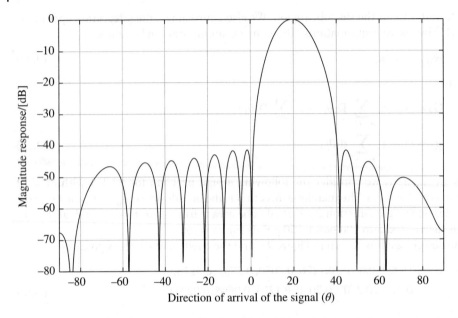

Figure 2.7 The resultant beam response for the off-broadside main beam design example using the least squares approach in (2.20).

For example, we can use the least squares design method to design a beamformer with its main beam pointing to the off-broadside direction $\theta = 20°$, while suppressing signals coming from directions $\theta \in [-90° \; 0°] \bigcup [40° \; 90°]$. The other parameters are the same as in the previous example.

The design result is shown in Figure 2.7 and the coefficient vector is given by

$$\mathbf{w} = [0.0248 - 0.0059i, 0.0240 - 0.0635i, -0.0829 - 0.1020i,$$
$$-0.2011 + 0.0320i, -0.0968 + 0.2498i, 0.1913 + 0.2392i,$$
$$0.3029 - 0.0455i, 0.0987 - 0.2491i, -0.1258 - 0.1602i,$$
$$-0.1301 + 0.0185i, -0.0256 + 0.0629i, 0.0142 + 0.0212i]^T \tag{2.23}$$

Instead of directly designing a beamformer pointing to the desired direction, we can design a beamformer with a broadside main beam and then add corresponding steering delays/phase shifts to the original design to obtain an equivalent beamformer with the desired main beam direction [50, 51]. Figure 2.8 shows such a general structure.

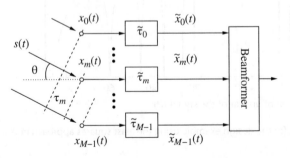

Figure 2.8 A general structure for beam steering.

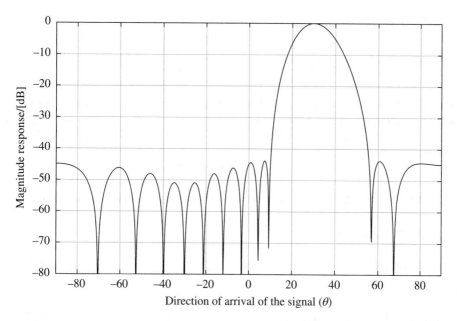

Figure 2.9 The new beam pattern after steering the result in (2.22) to 30°.

For a ULA with $d = \alpha \lambda / 2$, its beam response is given in Equation (2.14). For convenience, we rewrite it here

$$P(\theta) = \sum_{m=0}^{M-1} e^{-jma\pi \sin \theta} w_m^*$$ (2.24)

Suppose the set of coefficients w_m, $m = 0, 1, \ldots, M - 1$, has been designed to form a main beam pointing to the broadside ($\theta = 0$). To steer the beam to a new direction θ_0, we can add a delay of $\tilde{\tau}_0 = (M - 1)\frac{d \sin \theta_0}{c}$ to the first received array signal $x_0(t)$, a delay of $\tilde{\tau}_1 = (M - 2)\frac{d \sin \theta_0}{c}$ to the second received array signal $x_1(t)$, and so on. For a given signal frequency ω, this delay is simply a phase shift and the new response is given by

$$P(\sin \theta - \sin \theta_0) = e^{-j(M-1)a\pi \sin \theta_0} \sum_{m=0}^{M-1} w_m^* e^{-jma\pi(\sin \theta - \sin \theta_0)}$$

$$= e^{-j(M-1)a\pi \sin \theta_0} \sum_{m=0}^{M-1} \hat{w}_m^* e^{-jma\pi \sin \theta}$$ (2.25)

with $\hat{w}_m = w_m e^{-jma\pi \sin \theta_0}$.

Since $e^{-j(M-1)a\Omega \sin \theta_0}$ is a constant, the above equation represents a main beam in the direction of θ_0 and the equivalent beamformer coefficients for the new beam are given by $\hat{w}_m$, $m = 0, 1, \ldots, M - 1$.

As an example, we steer the main beam of the least squares based design result given in (2.22) from the broadside to $\theta_0 = 30°$ by multiplying w_m with $e^{-jm\pi \sin(\pi/6)} = e^{-jm(\pi/2)}$ and the new beam pattern is shown in Figure 2.9. Clearly the main beam has been steered to the new desired direction.

2.4   Adaptive Beamforming Algorithms

Instead of keeping a fixed beam response irrespective of the signal environments, we often need to change the beamformer's response according to the received data and accordingly the coefficients of the beamformer have to be adjusted constantly according to some adaptive algorithms. In this section, we introduce two classic adaptive beamformers: the reference signal based beamformer [3, 9, 30, 52, 53] and the Capon beamformer or, for the more general case, the linearly constrained minimum variance beamformer [9, 54, 55].

When we introduce these beamformers, we will consider their discrete-time versions. For convenience, the discrete-time version of the beamformer in Figure 2.1 is shown in Figure 2.10, where the beamformer output $y[n]$ at discrete-time index n is given by

$$y[n] = \sum_{m=0}^{M-1} w_m^* x_m[n]$$
$$= \mathbf{w}^H \mathbf{x}[n] \tag{2.26}$$

with

$$\mathbf{x}[n] = [x_0[n]\, x_1[n]\, \dots\, x_{M-1}[n]]^T \tag{2.27}$$

2.4.1   Reference Signal Based Beamformer

In some applications we could have a copy of the desired signal $s_0[n]$ and use it as the reference signal to adjust the beamformer coefficients using standard adaptive filtering algorithms, such as the least mean squares algorithm, the normalised least mean squares algorithm or the recursive least squares algorithm [9, 56].

A structure for the reference signal based beamformer is shown in Figure 2.11, where the M received sensor signals $x_0[n], x_1[n], \dots, x_{M-1}[n]$ are combined by the weight vector $\mathbf{w}$ to give the beamformer output $y[n]$ and its coefficients are adjusted by minimising a cost function based on the error signal $e[n]$ between the reference signal $r[n]$ and $y[n]$

$$e[n] = r[n] - y[n]$$
$$= r[n] - \mathbf{w}^H \mathbf{x}[n] \tag{2.28}$$

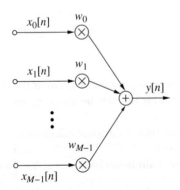

Figure 2.10 The discrete-time version of the beamformer in Figure 2.1.

Figure 2.11 The reference signal based beamforming structure.

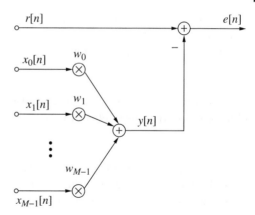

The least mean squares algorithm can be derived by minimising the following cost function

$$\xi = e[n]e^*[n]$$
$$= (r[n] - \mathbf{w}[n]^H \mathbf{x}[n])(r[n] - \mathbf{w}[n]^H \mathbf{x}[n])^*$$
$$= \hat{\sigma}_{rr}^2 - \mathbf{w}[n]^H \hat{\mathbf{p}} - \hat{\mathbf{p}}^H \mathbf{w}[n] + \mathbf{w}[n]^H \hat{\mathbf{R}}_{xx} \mathbf{w}[n] \tag{2.29}$$

where

$$\hat{\sigma}_{rr}^2 = |r[n]|^2$$
$$\hat{\mathbf{p}} = \mathbf{x}[n]d^*[n]$$
$$\hat{\mathbf{R}}_{xx} = \mathbf{x}[n]\mathbf{x}^H[n] \tag{2.30}$$

Note that we have included the time index n after $\mathbf{w}$ and use $\mathbf{w}[n]$ to indicate its value at the time instant n.

The gradient vector of the above cost function with respect to $\mathbf{w}^H$ can be evaluated as

$$\nabla \xi[n] = -\hat{\mathbf{p}} + \hat{\mathbf{R}}_{xx} \mathbf{w}[n]$$
$$= -e^*[n]\mathbf{x}[n] \tag{2.31}$$

Then, we can update the weight vector $\mathbf{w}[n]$ in the negative direction of the gradient with a step size μ, given by

$$\mathbf{w}[n+1] = \mathbf{w}[n] + \mu e^*[n]\mathbf{x}[n] \tag{2.32}$$

Instead of updating the weight vector with each new data sample coming in, we can also minimise the mean square error cost function $E\{e[n]e^*[n]\}$ as follows and derive the so-called Wiener solution [9, 56]

$$\tilde{\xi} = E\{e[n]e^*[n]\}$$
$$= E\{(r[n] - \mathbf{w}^H \mathbf{x}[n])(r[n] - \mathbf{w}^H \mathbf{x}[n])^*\}$$
$$= \sigma_{rr}^2 - \mathbf{w}^H \mathbf{p} - \mathbf{p}^H \mathbf{w} + \mathbf{w}^H \mathbf{R}_{xx} \mathbf{w} \tag{2.33}$$

where $E\{\cdot\}$ is the expectation operation and

$$\sigma_{rr}^2 = E\{r[n]r^*[n]\}$$
$$\mathbf{p} = E\{\mathbf{x}[n]d^*[n]\}$$
$$\mathbf{R}_{xx} = E\{\mathbf{x}[n]\mathbf{x}^H[n]\} \tag{2.34}$$

$\mathbf{p}$ is the cross-correlation vector between the reference signal and the received array signals, and $\mathbf{R}_{xx}$ is the covariance matrix of the received signals.

The gradient vector of the above cost function with respect to $\mathbf{w}^H$ can be evaluated as

$$\nabla \tilde{\xi} = -\mathbf{p} + \mathbf{R}_{xx}\mathbf{w} \tag{2.35}$$

The optimum solution or Wiener solution $\mathbf{w}_{opt}$ for the weight vector can then be obtained by setting the gradient to zero, which is given by

$$\mathbf{w}_{opt} = \mathbf{R}_{xx}^{-1}\mathbf{p} \tag{2.36}$$

In practice, we may not have the covariance matrix $\mathbf{R}_{xx}$ and the cross-correlation vector $\mathbf{p}$. In that case, we can use the following sample covariance matrix $\tilde{\mathbf{R}}_{xx}$ and sample cross-correlation vector to replace them

$$\tilde{\mathbf{R}}_{xx} = \frac{1}{K}\sum_{n=0}^{K-1}\mathbf{x}[n]\mathbf{x}^H[n] \ , \tilde{\mathbf{p}} = \frac{1}{K}\sum_{n=0}^{K-1}\mathbf{x}[n]d[n]^* \tag{2.37}$$

where K is the number of data samples received.

As an example, consider a ULA with $M = 6$ antennas. There is one desired signal arriving from the broadside and two interfering signals from DOA angles of $-20°$ and $30°$, respectively, with added zero-mean spatially and temporally white Gaussian noise. The signal-to-noise ratio (SNR) is 0 dB and the interference-to-noise ratio (INR) at each sensor is 20 dB. The total number of data samples K is 5000.

We use the Wiener solution in (2.36) to calculate the optimum solution for the weight vector and the result is given by

$$\mathbf{w} = [-0.0007 + 0.0011i, -0.0054 + 0.0043i, 0.0027 + 0.0098i,$$
$$-0.0068 - 0.0016i, 0.0079 + 0.0110i, -0.0041 - 0.0002i]^T \tag{2.38}$$

The resultant beam pattern is shown in Figure 2.12, where we can see the null clearly in the two interference directions.

2.4.2 The Capon Beamformer

If we do not have a reference signal, but know the direction of the signal of interest, then we can employ the so-called Capon beamformer to minimise the output power of the beamformer output, subject to a fixed unity response to the signal of interest [54]. One scenario where we know the direction of the signal of interest is when we perform beam scanning to detect potential targets of interest.

The output power of the beamformer is given by

$$E\{y[n]y^*[n]\} = E\{\mathbf{w}^H\mathbf{x}[n]\mathbf{x}^H[n]\mathbf{w}\}$$
$$= \mathbf{w}^H E\{\mathbf{x}[n]\mathbf{x}^H[n]\}\mathbf{w}$$
$$= \mathbf{w}^H \mathbf{R}_{xx}\mathbf{w} \tag{2.39}$$

Assuming the desired signal arrives from the direction θ_0, according to (2.3), to have a unity response to the desired signal, we must have

$$\mathbf{w}^H \mathbf{d}(\omega, \theta_0) = 1 \tag{2.40}$$

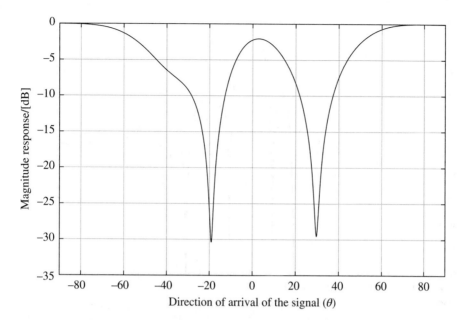

Figure 2.12 The resultant beam pattern for the reference signal based beamformer given in (2.38).

or simply

$$\mathbf{w}^H \mathbf{d}(\theta_0) = 1 \tag{2.41}$$

by removing the frequency ω in the expression.

Then, the Capon beamformer is formulated as

$$\mathbf{w}_{opt} = \arg \min_{\mathbf{w}} \ \mathbf{w}^H \mathbf{R}_{xx} \mathbf{w} \qquad \text{subject to} \qquad \mathbf{w}^H \mathbf{d}(\theta_0) = 1 \tag{2.42}$$

The optimum solution to the above constrained optimisation problem can be obtained by applying the Lagrange multipliers method [56, 57], and it is given by

$$\mathbf{w}_{opt} = \frac{\mathbf{R}_{xx}^{-1} \mathbf{d}(\theta_0)}{\mathbf{d}^H(\theta_0) \mathbf{R}_{xx}^{-1} \mathbf{d}(\theta_0)} \tag{2.43}$$

Again we can replace $\mathbf{R}_{xx}$ by the sample covariance matrix $\tilde{\mathbf{R}}_{xx}$ in (2.37) in practice.

The formulation in (2.42) only has one single constraint $\mathbf{w}[n]^H \mathbf{d}(\theta_0) = 1$ and we can add more constraints to have an enhanced control of the beamformer's response. For more details, please refer to [9].

As an example, consider a ULA with $M = 12$ antennas and all other conditions the same as in the reference signal based beamformer part. The optimum weight vector calculated by (2.43) is given by

$$\begin{aligned}
\mathbf{w} = [&0.0970 - 0.0039i, 0.0639 - 0.0066i, 0.0881 - 0.0104i, \\
&0.0910 - 0.0078i, 0.0712 + 0.0188i, 0.0933 - 0.0061i, \\
&0.1046 + 0.0039i, 0.0755 - 0.0055i, 0.0961 - 0.0098i, \\
&0.0698 + 0.0141i, 0.0740 - 0.0023i, 0.0754 + 0.0154i]^T
\end{aligned} \tag{2.44}$$

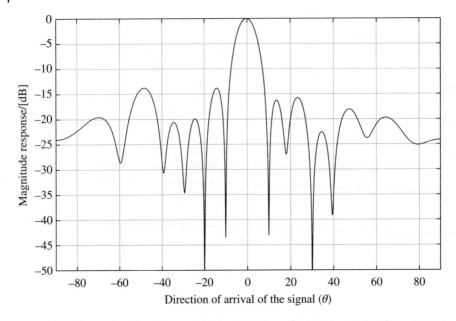

Figure 2.13 The resultant beam pattern for the Capon beamformer given in (2.43).

The resultant beam pattern is shown in Figure 2.13, where we can see that the beamformer has the desired unity response to the desired signal from the broadside and a null clearly in the two interference directions.

2.5 Blind Beamforming Algorithms

In this section we examine two additional adaptive beamforming algorithms which have specific applications in satellite navigation and wireless communications, respectively. A common feature of the two algorithms is that they do not need to know the direction of the desired signal or have a copy of the desired signal as the reference and therefore can achieve effective beamforming in a blind way. For this reason, they are called blind beamforming algorithms [9, 58–68].

2.5.1 The Power Minimisation Algorithm

In satellite-based navigation systems, such as the global positioning system (GPS) [69], the navigation signals received at the ground level have a very low power, typically 20–30 dB below the received thermal noise level [70]. Therefore, the performance of a satellite navigation system is extremely vulnerable to interferences from either intentional or unintentional sources. To tackle this problem, many interference suppression methods have been proposed in the past based on either the specific structure (such as cyclostationarity) or the DOA information of the signals, or both, employing traditional adaptive beamforming algorithms or some blind signal processing techniques [71–80].

Among them, the power minimisation method has a simple structure with a rather low complexity, but can effectively suppress the interfering signals given the very weak

Figure 2.14 A structure for the power minimisation based beamformer.

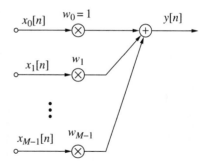

desired satellite signals [71, 74]. Since the power minimisation algorithm was proposed, some variations and extensions have been proposed with improved performance [77, 80]. Here we only focus on the basic form.

As shown in Figure 2.14, the first received signal $x_0[n]$ is the reference and w_m, $k = 1, \ldots, M - 1$ are the adaptive coefficients applied to the remaining received array signals. Since the interfering signals are much stronger than the desired signals and dominate the total received signals, when we minimise the power of the beamformer output $y[n]$, the interfering signals will be cancelled effectively, with mainly noise and the very weak desired signals left at the output.

In this context, we can consider that the first beamformer coefficient w_0 has a value of 1. Then the power minimisation problem can be formulated as

$$\mathbf{w}_{opt} = \arg \min_{\mathbf{w}} \mathbf{w}^H \mathbf{R}_{xx} \mathbf{w} \qquad \text{subject to} \qquad \mathbf{w}^H \mathbf{c} = 1 \tag{2.45}$$

where $\mathbf{c} = [1, 0, 0, \ldots]^T$ is an $M \times 1$ constraint vector and $\mathbf{w}$ is the coefficient vector as defined before.

Again, using the method of Lagrange multipliers, we can obtain the optimal solution as

$$\mathbf{w}_{opt} = \frac{\mathbf{R}_{xx}^{-1} \mathbf{c}}{\mathbf{c}^H \mathbf{R}_{xx}^{-1} \mathbf{c}} \tag{2.46}$$

As in the Capon beamformer case, in practice the covariance matrix $\mathbf{R}_{xx}$ of the received signals will be replaced by their sample covariance matrix $\tilde{\mathbf{R}}_{xx}$.

As an example, consider a ULA with $M = 6$ antennas. There is one very weak desired signal arriving from the DOA angle 10° and two strong interfering signals from DOA angles of −30° and 40°, respectively, with added spatially and temporally white zero-mean Gaussian noise. The SNR is −20 dB and the INR at each sensor is 20 dB. The total number of data samples K is 5000.

The optimum weight vector calculated by (2.46) is given by

$$\mathbf{w} = [1.0000 + 0.0000i, -0.0719 + 0.0205i, 0.2993 - 0.1803i,$$
$$0.1824 - 0.1854i, -0.1458 + 0.2150i, -0.1418 + 0.3458i]^T \tag{2.47}$$

The resultant beam pattern is shown in Figure 2.15, where we can see that the two interferences have been suppressed successfully.

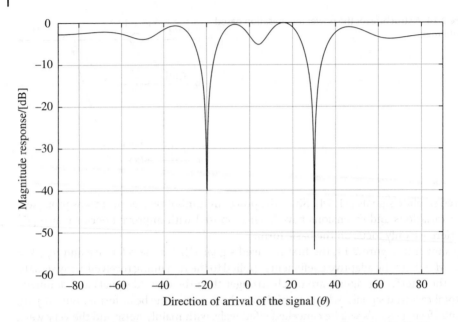

Figure 2.15 The resultant beam pattern for the power minimisation method given in (2.47).

2.5.2 The Constant Modulus Algorithm

In wireless communications, the transmitted signals often have some specific features inherent in the adopted modulation schemes, such as cyclostationarity, and these features can be exploited to perform effective beamforming without knowing the directions of the desired signals [81–85]. In the following, we introduce the CMA [3, 27, 61, 86–90], which is based on the constant modulus feature of the desired signals resulting from modulation schemes such as phase shift keying (PSK) or frequency shift keying.

There are many variations and the basic form is based on minimising the cost function

$$J_{CM} = E\{(|y[n]|^2 - \gamma)^2\} \tag{2.48}$$

where γ is the dispersion constant, defined by $\gamma = \frac{E\{|s_k|^4\}}{E\{|s_k|^2\}}$ with s_k being symbols of the modulation scheme.

Taking the gradient of J_{CM} with respect to the coefficient vector $\mathbf{w}^H$, we have

$$\nabla_{\mathbf{w}} J_{CM} = 2E\{|y[n]|^2 - \gamma)y^*[n]\mathbf{x}[n]\} \tag{2.49}$$

Using the instantaneous gradient $\hat{\nabla}_{\mathbf{w}} J_{CM}$ to replace $\nabla_{\mathbf{w}} J_{CM}$, we then obtain the updated equation for the constant modulus algorithm

$$\mathbf{w}[n+1] = \mathbf{w}[n] - \mu(\hat{\nabla}_{\mathbf{w}} J_{CM}) \tag{2.50}$$

$$= \mathbf{w}[n] - \mu(|y[n]|^2 - \gamma)y^*[n]\mathbf{x}[n] \tag{2.51}$$

where μ is the step size.

Now we consider an example for the CMA-based beamformer. A ULA with $M = 12$ antennas is employed with one desired quadrature PSK modulated signal from the broadside and two interfering signals from DOA angles of $-30°$ and $40°$, respectively. Both the SNR and the INR are 20 dB. The total number of data samples K is 5000. The

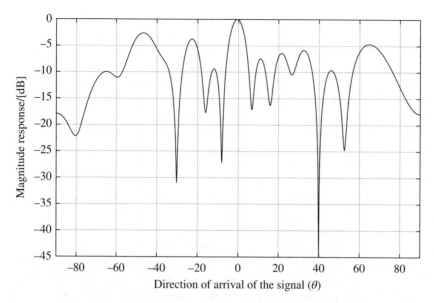

Figure 2.16 The resultant beam pattern for the CMA-based method given in (2.51).

stepsize $\mu = 0.00000001$ and the resultant weight vector after 5000 updates by (2.51) is given by

$$\mathbf{w} = [0.0714 - 0.0270i, 0.0148 + 0.0026i, -0.0025 + 0.0157i,$$
$$0.0140 - 0.0040i, 0.0320 + 0.0212i, 0.0024 - 0.0370i,$$
$$0.0395 + 0.0021i, 0.0072 + 0.0353i, 0.0031 + 0.0029i,$$
$$0.0353 + 0.0000i, 0.0157 + 0.0088i, 0.0603 - 0.0015i]^T \quad (2.52)$$

The resultant beam pattern is shown in Figure 2.16, where we can see that the two interferences have been suppressed successfully.

2.6 Low-cost Adaptive Beamforming

In this section we introduce two classes of beamforming methods which either have a low-cost implementation or enable a low-cost implementation.

2.6.1 Analogue and Digital Hybrid Beamforming

Nowadays various array signal processing tasks such as beamforming are increasingly being implemented digitally. For digital processing, the first step is to employ an ADC to transform the received analogue antenna signals into the digital format. This can be performed at either the RF or the baseband. According to the Nyquist sampling theorem [40], the minimum sampling frequency required to avoid aliasing is twice the highest frequency of the analogue signal. Therefore, to reduce the sampling frequency required and therefore the cost and power consumption of the ADC circuits, the ADC is preferably applied to the baseband signal.

In the past few decades we have seen an explosive development in wireless or mobile communications technology and the next-generation (the fifth-generation, 5G) communication systems will be deployed soon. Two of the key enabling technologies for 5G are massive MIMO and millimetre-wave communication [91]. Both require the employment of a large number of antennas (i.e. hundreds of them) working at high frequencies with a very wide bandwidth. If we implement the beamforming process in a fully digital manner, i.e. we convert all the received analogue antenna signals into digital ones satisfying the Nyquist sampling constraint before performing any beamforming operations, we would need at least the same number of high-speed ADCs as the antenna number (or twice the antenna number for complex baseband signals). For the digital beamforming algorithms to work effectively, we also have to make sure the large number of ADCs are fully synchronised. The extremely high cost associated with the ADCs themselves and the high-level power consumption will render the digital approach practically infeasible. There are two possible approaches to solve this problem.

One is to reduce the required dynamic range of the ADCs so that power consumption can be reduced. This can be achieved by employing some analogue beamforming techniques [92] or some judicially designed transformations [93, 94]. The other approach is to employ a hybrid beamforming structure [95–97], where beamforming is performed in the analogue domain first to reduce the number of analogue channels, which are then converted to digital via a reduced number of ADCs, and after that traditional digital beamforming can then be performed.

Various hybrid beamforming structures have been proposed in the past and one of them is the subaperture based hybrid beamformer [97], which can be further divided into two types. One type is called localised configuration, and an example is shown in Figure 2.17.

In this example, there are in total KM antennas and they are split into K groups, with M antennas for each group. The groups of antennas are located next to each other and there is no spatial overlap between any two groups. The kth group of antenna received signals $x_{k,m}(t)$, $m = 0, 1, \cdots, M - 1$ are first processed by an analogue beamformer with analogue coefficients $w_{k,m}$, $m = 0, 1, \cdots, M - 1$ and the output of the beamformer is denoted by $x_k(t)$; $x_k(t)$ is then converted into digital format $x_k[n]$ by an ADC. All of the K digital signals $x_k[n]$, $k = 0, 1, \cdots, K - 1$ are processed by a digital beamformer with digital coefficients w_k, $k = 0, 1, \cdots, K - 1$ to give the final output $y[n]$.

The other type is called the interleaved configuration, as shown in Figure 2.18. In this case, the KM antennas are split into M groups, with K antennas for each group. The groups of antennas are located next to each other and there is no overlap spatially between any two groups. To form the first analogue beamformer with analogue coefficients $w_{m,0}$, $m = 0, 1, \cdots, M - 1$, the first signal $x_{0,0}(t)$ is taken from the first antenna received signal from the first group, the second signal $x_{1,0}(t)$ is taken from the first antenna received signal from the second group, and so on. For the kth analogue beamformer with analogue coefficients $w_{m,k}$, $m = 0, 1, \cdots, M - 1$, the first signal $x_{0,k}(t)$ is taken from the kth antenna received signal from the first group, the second signal $x_{1,k}(t)$ is taken from the kth antenna received signal from the second group, and so on. In this way, we still have K analogue beamformers, which give K analogue outputs $x_k(t)$, $k = 0, 1, \cdots, K - 1$; as in the first example they are converted into digital signals, which are further processed by a digital beamformer to give the final output $y[n]$.

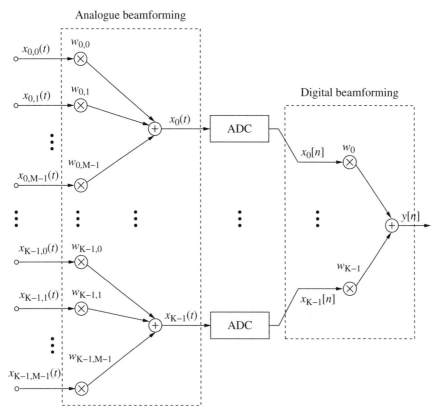

Figure 2.17 The subarray-based hybrid beamforming structure with localised configuration.

Although this hybrid structure can reduce the cost, its solution is not globally optimal and one key topic of research is to find the combination of analogue and digital coefficients to give the best possible overall beamforming performance.

2.6.2 Robust Adaptive Beamforming

Apart from the issues related to the ADC, another challenge for large-scale antenna arrays is calibration. As we have seen already, in adaptive beamforming and also in the direction of arrival estimation, in most cases we need to know the exact form of the array steering vector, i.e. we need to know the exact radiation pattern of each antenna and also their spatial locations. Given the mutual coupling problem, we also need to model the mutual coupling effect between antennas in a fairly accurate manner. These are already very difficult problems to solve even for a small antenna array system, not to mention a system with hundreds of antennas. Instead of trying to find the exact steering vector of an antenna array directly, another way is to develop array signal processing algorithms which are robust against various steering vector errors. Beamforming research falling into this area is called robust beamforming [6, 9, 98].

Two representative robust adaptive beamforming techniques are diagonal loading and worst-case based formulation [15, 99–106]. By reducing the workload on calibration, we

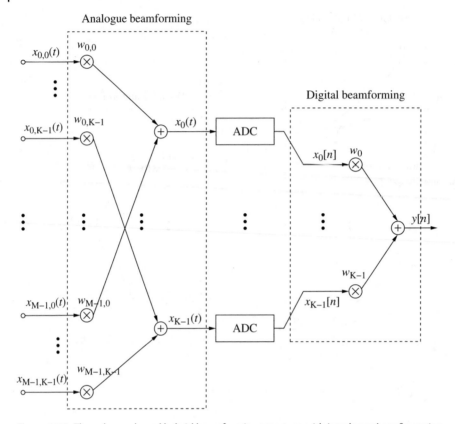

Figure 2.18 The subarray based hybrid beamforming structure with interleaved configuration.

can significantly reduce the cost of implementing a large antenna array system. More-over, traditionally we normally assume the array antennas have the same radiation pattern; however, it is very costly to maintain this standard for a large number of antennas located at different spatial positions and there are always some variations in their radiation pattern among them. This inconsistency in their response can be treated as a steering vector error too and robust beamforming algorithms provide a way to reduce this consistency requirement, allowing further reduction in the cost of building large antenna array systems.

In the following, we first examine the diagonal loading based technique in detail.

The diagonal loading method modifies the received signal covariance matrix $\mathbf{R}_{xx}$ by adding a small value δ to its diagonal elements as follows

$$\overline{\mathbf{R}} = \mathbf{R}_{xx} + \delta \mathbf{I} \tag{2.53}$$

where $\mathbf{I}$ is the identity matrix and δ is the diagonal loading factor and needs to be adjusted according to the specific scenario, which is the major drawback of this method.

With this modified covariance matrix, the Capon beamforming problem changes to

$$\mathbf{w} = \arg\min_{\mathbf{w}} \ \mathbf{w}^H (\mathbf{R}_{xx} + \delta \mathbf{I}) \mathbf{w} \qquad \text{subject to} \qquad \mathbf{w}^H \mathbf{d}(\theta_0) = 1 \tag{2.54}$$

or

$$\mathbf{w} = \arg \min_{\mathbf{w}} \mathbf{w}^H \mathbf{R}_{xx} \mathbf{w} + \delta \mathbf{w}^H \mathbf{w} \qquad \text{subject to} \qquad \mathbf{w}^H \mathbf{d}(\theta_0) = 1 \qquad (2.55)$$

As given in Equation (2.43), the new optimum solution $\mathbf{w}_{opt}$ to this modified problem becomes

$$\mathbf{w}_{opt} = \frac{\overline{\mathbf{R}}_{xx}^{-1} \mathbf{d}(\theta_0)}{\mathbf{d}^H(\theta_0) \overline{\mathbf{R}}_{xx}^{-1} \mathbf{d}(\theta_0)} \qquad (2.56)$$

From Equation (2.55), we can see that the diagonal loading method can be considered as adding a penalty term based on the norm of the weight vector $\mathbf{w}^H \mathbf{w}$ to the original cost function and therefore is a norm-constrained approach to robust adaptive beamforming.

To provide some further insight into why this approach can provide a robust performance against steering vector errors, we consider a simple scenario where, due to miscalibration of the array, the steering vector error in $\mathbf{d}(\theta_0)$ of the desired signal takes the form of a small angle error ε, i.e. the assumed direction of arrival of the desired signal is θ_0, while its real direction is $\theta_0 + \varepsilon$.

If we use the standard Capon beamformer, it will form a unity response to the direction θ_0 and suppress signals coming from any other directions, including $\theta_0 + \varepsilon$, which means the desired signal will be suppressed as interference in the beamformer output.

To understand how the norm constraint on $\mathbf{w}$ works, we first obtain the derivative of the squared response $|P(\theta)|^2$ with respect to the DOA angle θ as follows

$$\frac{d(|P(\theta)|^2)}{d\theta} = \frac{d(\mathbf{w}^H \mathbf{d}(\theta))}{d\theta} \mathbf{d}^H(\theta)\mathbf{w} + \mathbf{w}^H \mathbf{d}(\theta)\frac{d(\mathbf{d}^H(\theta)\mathbf{w})}{d\theta}$$

$$= \frac{d(\mathbf{w}^H \mathbf{d}(\theta))}{d\theta} \mathbf{d}^H(\theta)\mathbf{w} + \left(\frac{d(\mathbf{w}^H \mathbf{d}(\theta))}{d\theta} \mathbf{d}^H(\theta)\mathbf{w}\right)^H$$

$$= 2\Re\left(\mathbf{w}^H \frac{d(\mathbf{d}(\theta))}{d\theta} \mathbf{d}^H(\theta)\mathbf{w}\right) \qquad (2.57)$$

where $\Re\{\}$ denotes the real part of its complex-valued variable.

Since the signal of interest is suppressed at the beamformer output, the resultant beam pattern of the Capon beamformer at $\theta_0 + \varepsilon$ will have a response very close to zero. On the other hand, due to the constraint $\mathbf{w}^H \mathbf{d}(\theta_0) = 1$, the response of the beamformer will change from unity at θ_0 to almost zero at $\theta_0 + \varepsilon$, leading to a very large value of the derivative $\frac{d(|P(\theta)|^2)}{d\theta}$ at $\theta = \theta_0$. Since both $\frac{d(\mathbf{d}(\theta))}{d\theta}$ and $\mathbf{d}(\theta)$ are bounded in Equation (2.57), to have a very large derivative value at θ_0 we have to increase the norm of the weight vector significantly. By constraining the norm of the weight vector through diagonal loading, we can reduce this derivative value at θ_0 and therefore avoid having an almost zero response at $\theta_0 + \varepsilon$. As a result, the desired signal will not be suppressed in the diagonal loading based method as significantly as in the standard Capon beamformer.

As an example, we consider a ULA with $M = 12$ antennas. There is one desired signal arriving from $3°$ and two interfering signals from DOA angles of $-20°$ and $30°$, respectively, with added zero-mean spatially and temporally white Gaussian noise. The SNR is 0 dB and the INR at each antenna is 20 dB. The total number of data samples is 5000. We assume the desired signal arrives from the broadside so that a $3°$ error is present in the assumed steering vector.

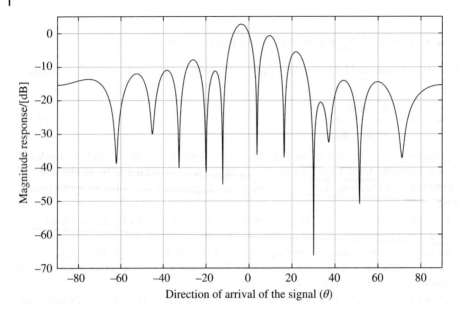

Figure 2.19 The resultant beam pattern for the Capon beamformer given in (2.43) with 3° steering vector error.

First, we apply the standard Capon beamformer to the scenario and the resultant optimum weight vector is given by

$$\mathbf{w} = [-0.0997 - 0.8717i, -0.0404 - 0.7779i, -0.2036 - 0.6110i,$$
$$-0.2681 - 0.4257i, -0.0782 - 0.3218i, 0.2284 - 0.3613i,$$
$$0.3020 - 0.2787i, 0.3005 - 0.0257i, 0.4877 + 0.1708i,$$
$$0.6730 + 0.0832i, 0.7291 - 0.1845i, 0.8820 - 0.0854i]^T \qquad (2.58)$$

and its beam pattern is shown in Figure 2.19, where we can see that the beamformer has the desired unity response at the broadside and a null at the two interference directions. Moreover, it also has a null at the real direction $\theta = 3°$ of the desired signal as in this case the desired signal has been treated as an interference by the Capon beamformer. We can also observe the steep fall from unity at $\theta = 0°$ to almost zero at $\theta = 3°$ as predicted in the earlier discussion.

To avoid suppressing the desired signal, we can apply the diagonal loading based method given in Equation (2.56). With a diagonal loading factor $\delta = 20$, the resultant weight vector is given by

$$\mathbf{w} = [0.0496 - 0.1166i, 0.0548 - 0.1118i, 0.0492 - 0.1080i,$$
$$0.0472 - 0.0967i, 0.0559 - 0.0883i, 0.0679 - 0.0881i,$$
$$0.0704 - 0.0862i, 0.0723 - 0.0762i, 0.0830 - 0.0677i,$$
$$0.0932 - 0.0730i, 0.0953 - 0.0801i, 0.1014 - 0.0759i]^T \qquad (2.59)$$

and its beam pattern is shown in Figure 2.20, where we can see that the null has moved away from the desired signal direction $\theta = 3°$ and the fall from $\theta = 0°$ to $\theta = 3°$ has

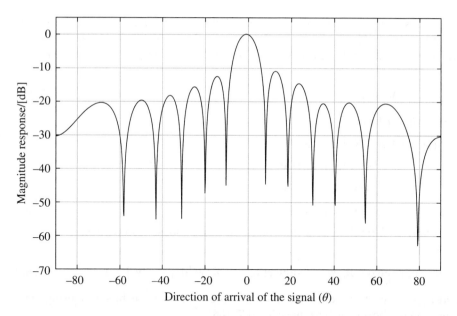

Figure 2.20 The resultant beam pattern for the diagonal loading based Capon beamformer given in (2.56) with 3° steering vector error.

been reduced significantly, demonstrating the effectiveness of the diagonal loading based robust adaptive beamforming method.

Next we briefly introduce the worst-case based method for robust adaptive beamforming. First, we denote the actual steering vector of the desired signal as $\mathbf{d}_0$, which belongs to a spherical set, given by

$$\mathcal{D}_0 = \{\mathbf{d}_0 \mid \mathbf{d}_0 = \overline{\mathbf{d}}_0 + \mathbf{e}, \; \|\mathbf{e}\| \leq \varepsilon\} \tag{2.60}$$

where $\overline{\mathbf{d}}_0$ is the assumed/ideal steering vector, $\mathbf{e}$ is the norm-constrained error vector between the assumed and the real steering vector, and ε is the maximum norm value of the error vector.

Following the Capon beamformer idea, we can minimise the power of the beamformer output subject to the constraint that the minimum magnitude value of the array response for all the steering vectors $\mathbf{d}_0$ belonging to the above spherical set is greater than one, i.e.

$$\mathbf{w} = \arg \min_{\mathbf{w}} \; \mathbf{w}^H \mathbf{R}_{xx} \mathbf{w} \qquad \text{subject to} \qquad \min_{\mathbf{d}_0 \in \mathcal{D}_0} |\mathbf{w}^H \mathbf{d}_0| \geq 1 \tag{2.61}$$

Since the response of the beamformer in the worst case for the whole set $\mathcal{D}_0$ including the real steering vector larger than one, the resultant beamformer will not suppress the desired signal due to the steering vector error.

To solve the above problem, we can rewrite the constraint by constraining its real component to be larger than one and its imaginary component to be zero, since a phase shift to the beamformer output will not change the power of the signal [101]

$$\min_{\mathbf{d}_0 \in \mathcal{D}_0} \; \mathfrak{R}\{\mathbf{w}^H \mathbf{d}_0\} \geq 1,$$

$$\mathfrak{I}\{\mathbf{w}^H \mathbf{d}_0\} = 0 \tag{2.62}$$

where $\mathfrak{I}\{\}$ denotes the imaginary part of its complex-valued variable.

By maximising the real part of a complex number using a phase shift, its imaginary part will be automatically minimised, therefore we can remove the constraint on the imaginary part and simplify the constraint in (2.62) into a single real component constraint,

$$\min_{\mathbf{d}_0 \in D_0} \Re\{\mathbf{w}^H \mathbf{d}_0\} \geq 1 \tag{2.63}$$

From (2.60), we have

$$
\begin{aligned}
\Re\{\mathbf{w}^H \mathbf{d}_0\} &= \Re\{\mathbf{w}^H(\overline{\mathbf{d}}_0 + \mathbf{e})\} \\
&= \Re\{\mathbf{w}^H \overline{\mathbf{d}}_0\} + \Re\{\mathbf{w}^H \mathbf{e}\} \\
&\geq \Re\{\mathbf{w}^H \overline{\mathbf{d}}_0\} - |\mathbf{w}^H \mathbf{e}| \\
&\geq \Re\{\mathbf{w}^H \overline{\mathbf{d}}_0\} - \varepsilon\|\mathbf{w}\|
\end{aligned}
\tag{2.64}
$$

As a result, the formulation for the worst-case based robust adaptive beamformer can be changed to

$$\mathbf{w} = \arg\min_{\mathbf{w}} \mathbf{w}^H \mathbf{R}_{xx} \mathbf{w} \qquad \text{subject to} \qquad \Re\{\mathbf{w}^H \overline{\mathbf{d}}_0\} - \varepsilon\|\mathbf{w}\| \geq 1 \tag{2.65}$$

There is no closed-form solution to the above problem, but it can be solved using some optimisation toolboxes, such as those in [107, 108].

2.7 Summary of the Chapter

In this chapter some basic digital beamforming methods and algorithms have been introduced in detail, including beam steering and two fixed beamformer design methods, two traditional adaptive beamforming algorithms, two blind adaptive beamforming algorithms with specific applications in satellite navigation and wireless communications, and two low-cost beamforming approaches for next-generation wireless communication systems. In particular, for low-cost beamforming, the hybrid beamforming structure employing analogue and digital beamformers at different stages was reviewed, followed by two robust adaptive beamforming algorithms, which allow a low-cost implementation due to its tolerance to various errors in the system.

References

1 A.J. Paulraj and C.B. Papadias. Space-time processing for wireless communications. *IEEE Signal Processing Magazine*, 14(6):49–83, Nov 1997.

2 L.C. Godara. Application of antenna arrays to mobile communications, Part I: Performance improvement, feasibility, and system considerations. *Proceedings of the IEEE*, 85(7):1031–1060, July 1997.

3 L.C. Godara. Application of antenna arrays to mobile communications, Part II: Beam-forming and direction-of-arrival estimation. *Proceedings of the IEEE*, 85(8):1195–1245, August 1997.

4 G.V. Tsoulos. Smart antennas for mobile communication systems: Benefits and challenges. *IEE Electronics & Communications Engineering Journal*, 11(2):84–94, April 1999.

5 N. Fourikis. *Advanced Array Systems, Applications and RF Technologies*. Academic Press, London, 2000.

6 H.L. Van Trees. *Optimum Array Processing, Part IV of Detection, Estimation, and Modulation Theory*. Wiley, New York, 2002.

7 R.A. Monzingo and T.W. Miller. *Introduction to Adaptive Arrays*. SciTech Publishing Inc., Raleigh, NC, 2004.

8 B. Allen and M. Ghavami. *Adaptive Array Systems, Fundamentals and Applications*. John Wiley & Sons, Chichester, 2005.

9 W. Liu and S. Weiss. *Wideband Beamforming: Concepts and Techniques*. John Wiley & Sons, Chichester, 2010.

10 A. Moffet. Minimum-redundancy linear arrays. *IEEE Transactions on Antennas and Propagation*, 16(2):172–175, March 1968.

11 R.T. Hoctor and S.A. Kassam. The unifying role of the coarray in aperture synthesis for coherent and incoherent imaging. *Proceedings of the IEEE*, 78(4):735–752, April 1990.

12 P. Pal and P.P. Vaidyanathan. Nested arrays: a novel approch to array processing with enhanced degrees of freedom. *IEEE Transactions on Signal Processing*, 58(8):4167–4181, August 2010.

13 P.P. Vaidyanathan and P. Pal. Sparse sensing with co-prime samplers and arrays. *IEEE Transactions on Signal Processing*, 59(2):573–586, Feb. 2011.

14 M.B. Hawes and W. Liu. Location optimisation of robust sparse antenna arrays with physical size constraint. *IEEE Antennas and Wireless Propagation Letters*, 11:1303–1306, November 2012.

15 M.B. Hawes and W. Liu. Compressive sensing based approach to the design of linear robust sparse antenna arrays with physical size constraint. *IET Microwaves, Antennas & Propagation*, 8:736–746, July 2014.

16 M.B. Hawes and W. Liu. Sparse array design for wideband beamforming with reduced complexity in tapped delay-lines. *IEEE Transactions on Audio, Speech and Language Processing*, 22:1236–1247, August 2014.

17 Q. Shen, W. Liu, W. Cui, S.L. Wu, Y.D. Zhang, and M. Amin. Low-complexity direction-of-arrival estimation based on wideband co-prime arrays. *IEEE Transactions on Audio, Speech and Language Processing*, 23:1445–1456, September 2015.

18 Q. Shen, W. Liu, W. Cui, and S.L. Wu. Extension of co-prime arrays based on the fourth-order difference co-array concept. *IEEE Signal Processing Letters*, 23(5):615–619, May 2016.

19 J. Cai, W. Liu, R. Zong, and Q. Shen. An expanding and shift scheme for constructing fourth-order difference co-arrays. *IEEE Signal Processing Letters*, 24(4):480–484, April 2017.

20 R.T. Compton. The tripole antenna: An adaptive array with full polarization flexibility. *IEEE Transactions on Antennas and Propagation*, 29(6):944–952, November 1981.

21 A. Nehorai, K.C. Ho, and B.T.G. Tan. Minimum-noise-variance beamformer with an electromagnetic vector sensor. *IEEE Transactions on Signal Processing*, 47(3):601–618, March 1999.

22 M.D. Zoltowski and K.T. Wong. ESPRIT-based 2D direction finding with a sparse uniform array of electromagnetic vector-sensors. *IEEE Transactions on Signal Processing*, 48:2195–2204, August 2000.

23 X.R. Zhang, Z.W. Liu, Y.G. Xu, and W. Liu. Adaptive tensorial beamformer based on electromagnetic vector-sensor arrays with coherent interferences. *Multidimensional Systems and Signal Processing*, 26:803–821, July 2015.

24 M.B. Hawes and W. Liu. Design of fixed beamformers based on vector-sensor arrays. *International Journal of Antennas and Propagation*, 2015, 2015.

25 M.B. Hawes, W. Liu, and L. Mihaylova. Compressive sensing based design of sparse tripole arrays. *Sensors*, 15(12):31056–31068, 2015.

26 M.D. Jiang, W. Liu, and Y. Li. Adaptive beamforming for vector-sensor arrays based on reweighted zero-attracting quaternion-valued LMS algorithm. *IEEE Transactions on Circuits and Systems II: Express Briefs*, 63:274–278, March 2016.

27 W. Liu. Channel equalization and beamforming for quaternion-valued wireless communication systems. *Journal of the Franklin Institute*, 354:8721–8733, December 2017.

28 X. Lan and W. Liu. Fully quaternion-valued adaptive beamforming based on crossed-dipole arrays. *Electronics*, 6(2):34(1–16), 2017.

29 B. Sklar. *Digital Communications: Fundamentals and Applications*. Prentice Hall, New Jersey, 2001.

30 B.D. Van Veen and K.M. Buckley. Beamforming: a versatile approach to spatial filtering. *IEEE Acoustics, Speech, and Signal Processing Magazine*, 5(2):4–24, April 1988.

31 Y. Kamp and J. Thiran. Chebyshev approximation for two-dimensional nonrecursive digital filters. *IEEE Transactions on Circuits and Systems*, CAS-22(3):208–218, March 1975.

32 S. Nordebo, I. Claesson, and S. Nordholm. Weighted Chebyshev approximation for the design of broadbandbeamformers using quadratic programming. *IEEE Signal Processing Letters*, 1:103–105, July 1994.

33 S.E. Nordholm, V. Rehbock, K.L. Tee, and S. Nordebo. Chebyshev optimization for the design of broadband beamformers in the near field. *IEEE Transactions on Circuits and Systems II: Analog and Digital Signal Processing*, 45(1):141–143, January 1998.

34 H. Lebret and S. Boyd. Antenna array pattern synthesis via convex optimization. *IEEE Transactions on Signal Processing*, 45(3):526–532, March 1997.

35 D.P. Scholnik and J.O. Coleman. Optimal design of wideband array patterns. In *Proceedings of the IEEE International Radar Conference*, pages 172–177, Washington, US, May 2000.

36 D.P. Scholnik and J.O. Coleman. Formulating wideband array-pattern optimizations. In *Proceedings of the IEEE International Conference on Phased Array Systems and Technology*, pages 489–492, Dana Point, California, May 2000.

37 D.P. Scholnik and J.O. Coleman. Superdirectivity and SNR constraints in wideband array-pattern design. In *Proceedings of the IEEE International Radar Conference*, pages 181–186, Atlanta, US, May 2001.

38 H. Duan, B.P. Ng, C.M. See, and J. Fang. Applications of the SRV constraint in broadband pattern synthesis. *Signal Processing*, 88:1035–1045, April 2008.

39 Y. Zhao, W. Liu, and R.J. Langley. Efficient design of frequency invariant beam-formers with sensor delay-lines. In *Proceedings of the IEEE Workshop on Sensor Array and Multichannel Signal Processing*, pages 335–339, Darmstadt, Germany, July 2008.

40 A.V. Oppenheim and R.W. Schafer. *Digital Signal Processing*. Prentice-Hall, Englewood Cliffs, 1975.

41 The MathWorks, Inc., Natick, MA. *MATLAB 2016b*, 2016.

42 C.L. Lawson and R.J. Hanson. *Solving Least Squares Problems*. Automatic Computation. Prentice Hall, Englewood Cliffs, NJ, 1974.

43 A. Björck. *Numerical Methods for Least Squares Problems*. Society for Industrial and Applied Mathematics (SIAM), Philadelphia, PA, 1996.

44 M. Lang, I.W. Selesnick, and C.S. Burrus. Constrained least squares design of 2D FIR filters. *IEEE Transactions on Signal Processing*, 44:1234–124,May 1996.

45 S.C. Pei and C.C. Tseng. A new eigenfilter based on total least squares error criterion. *IEEE Transactions on Circuits & Systems I: Regular Papers*, 48:699–709, 2001.

46 S. Doclo and M. Moonen. Design of far-field and near-field broadband beamformers using eigenfilters. *Signal Processing*, 83(12):2641–2673, December 2003.

47 Y. Zhao, W. Liu, and R.J. Langley. A least squares approach to the design of frequency invariant beamformers. In *Proceedings of the European Signal Processing Conference*, pages 844–848, Glasgow, Scotland, August 2009.

48 Y. Zhao, W. Liu, and R.J. Langley. Subband design of fixed wideband beamformers based on the least squares approach. *Signal Processing*, 91:1060–1065, April 2011.

49 Y. Zhao, W. Liu, and R.J. Langley. An application of the least squares approach to fixed beamformer design with frequency invariant constraints. *IET Signal Processing*, pages 281–291,June 2011.

50 W. Liu and S. Weiss. Beam steering for wideband arrays. *Signal Processing*, 89:941–945, May 2009.

51 W. Liu and D.R. Morgan. A spatial filtering approach to electronic wideband beam steering. *In Proceedings of the SPIE Defence, Security, and Sensing Symposium*, volume 8061, pages 80610G1–80610G9, Orlando, USA, April 2011.

52 W. Liu. Adaptive wideband beamforming with sensor delay-lines. *Signal Processing*, 89:876–882, May 2009.

53 N. Lin, W. Liu, and R.J. Langley. Performance analysis of an adaptive broadband beamformer based on a two-element linear array with sensor delay-line processing. *Signal Processing*, 90:269–281, January 2010.

54 J. Capon. High-resolution frequency-wavenumber spectrum analysis. *Proceedings of the IEEE*, 57(8):1408–1418, August 1969.

55 O.L. Frost, III. An algorithm for linearly constrained adaptive array processing. *Proceedings of the IEEE*, 60(8):926–935, August 1972.

56 S. Haykin. *Adaptive Filter Theory*. Prentice Hall, Englewood Cliffs, New York, 3rd edition, 1996.

57 D.H. Johnson and D.E. Dudgeon. *Array Signal Processing: Concepts and Techniques*. *Signal Processing Series*. Prentice Hall, Englewood Cliffs, NJ, 1993.

58 J.F. Cardoso and A. Souloumiac. Blind beamforming for non-Gaussian signals. *IEE Proceedings F, Radar and Signal Processing*, 140:362–370,December 1993.

59 Gönen, E. and Mendel, J.M. Applications of cumulants to array processing. III. Blind beamforming for coherent signals. *IEEE Transactions on Signal Processing*, 45:2252–2264, September 1997.

60 J. Sheinvald. On blind beamforming for multiple non-gaussian signals and the constant-modulus algorithm. *IEEE Transactions on Signal Processing*, 46:1878–1885, July 1998.

61 Z. Ding and Y. Li. *Blind Equalisation and Identification*. Signal Processing and Communications. CRC, New York, 2001.

62 K. Yang, T. Ohira, Y. Zhang, and C.Y. Chi. Super-exponential blind adaptive beamforming. *IEEE Transactions on Signal Processing*, 52(6):1549–1563, June 2004.

63 W. Liu and D.P. Mandic. Semi-blind source separation for convolutive mixtures based on frequency invariant transformation. In *Proceedings of the IEEE International Conference on Acoustics, Speech, and Signal Processing*, volume 5, pages 285–288, Philadelphia, USA, March 2005.

64 X. Huang, H.C. Wu, and J.E. Principe. Robust blind beamforming algorithm using joint multiple matrix diagonalization. *IEEE Sensors Journal*, 7:130–136, January 2007.

65 S. Chen, W. Yao, and L. Hanzo. Semi-blind adaptive spatial equalisation for MIMO systems with high-order QAM signalling. *IEEE Transactions on Wireless Communications*, 7(11):4486–4491, November 2008.

66 W. Liu. Blind adaptive wideband beamforming for circular arrays based on phase mode transformation. *Digital Signal Processing*, 21:239–247, March 2011.

67 L. Zhang, W. Liu, and R.J. Langley. Low-complexity constant modulus algorithms for blind beamforming based on symmetrically distributed arrays. In *Proceedings of the IEEE Workshop on Statistical Signal Processing*, pages 385–388, Nice, France, June 2011.

68 W. Liu. Wideband beamforming for multi-path signals based on frequency invariant transformation. *International Journal of Automation and Computing*, 9:420–428, August 2012.

69 E.D. Kaplan and C. Hegarty. *Understanding GPS: Principles and Applications*, 2nd edition. Artech House, UK, 2005.

70 P.T. Capozza, B.J. Holland, T.M. Hopkinson, and R.L. Landrau. A single-chip narrow-band frequency-domain excisor for a global positioning system receiver. *IEEE Journal of Solid-State Circuits*, 35:401–411, May 2000.

71 M.D. Zoltowski and A.S. Gecan. Advanced adaptive null steering concepts for GPS. *Military Communications Conference*, pages 1214–1218, November 1995.

72 G.F. Hatke. Adaptive array processing for wideband nulling in GPS system. In *Proceedings of the 32nd Asilomar Conference on Signals Systems and Computers*, pages 1332–1336, 1998.

73 R.L. Fante and J.J. Vaccaro. Wideband cancellation of interference in a GPS receiver array. *IEEE Transactions on Aerospace and Electronic Systems*, 36(2):549–564, April 2000.

74 R.S. Jay. Interference mitigation approaches for the global positioning system. *Lincoln Laboratory Journal*, pages 167–180, 2003.

75 W. Sun and M.G. Amin. A novel interference suppression scheme for global navigation satellite systems using antenna array. *IEEE Journal on Selected Areas in Communications*, 36:999–1012, May 2005.

76 D. Lu, Q. Feng, and R.B. Wu. Survey on interference mitigation via adaptive array processing in GPS. *Proceedings of the Progress in Electronmagnetics Research Symposium*, 2(4):357–362, March 2006.

77 W. Huang, R.B. Wu, and D. Lu. A novel blind GPS anti-jamming algorithm based on subspace technique. In *Proceedings of the International Conference on Signal Processing*, Beijing, China, 2006.

78 P. Li, D. Lu, R.B. Wu, and Z.G. Su. Adaptive anti-jamming algorithm based on the characteristics of the GPS signal. In *Proceedings of the International Symposium on Intelligent Signal Processing and Communication Systems*, pages 192–195, Tianjin, China, December 2007.

79 H. Yao. A reduced-rank stap method based on solution of linear equation. In *Proceedings of the International Conference on Computer Design and Applications*, volume 1, pages 235–238, Octorber 2010.

80 B. Qiu, W. Liu, and R.B. Wu. Blind interference suppression for satellite navigation signals based on antenna arrays. In *Proceedings of the IEEE China Summit and International Conference on Signal and Information Processing*, pages 370–373, July 2013.

81 B.G. Agee, S.V. Schell, and W.A. Gardner. Spectral self-coherence restoral: a new approach to blind adaptive signal extraction using antenna arrays. *Proceedings of the IEEE*, 78(4):753–767, 1990.

82 W.A. Gardner. Exploitation of spectral redundancy in cyclostationary signals. *IEEE Signal Processing Magazine*, 8(2):14–36, April 1991.

83 Q. Wu and K.M. Wong. Blind adaptive beamforming for cyclostationary signals. *IEEE Transactions on Signal Processing*, 44(11):2757–2767, Nov. 1996.

84 K.L. Du and M. Swamy. A class of adaptive cyclostationary beamforming algorithms. *Circuits, Systems, and Signal Processing*, 27:35–63, 2008.

85 W. Zhang and W. Liu. Low-complexity blind beamforming based on cyclostationarity. In *Proceedings of the European Signal Processing Conference*, pages 1349–1353, Bucharest, *Romania*,August 2012.

86 R. Gooch and J. Lundell. CM array: an adaptive beamformer for constant modulus signals. In *Proceedings of the IEEE International Conference on Acoustics, Speech, and Signal Processing*, pages 2523–2526, New York, NY, 1986.

87 H. Krim and M. Viberg. Two decades of array signal processing research: the parametric approach. *IEEE Signal Processing Magazine*, 13(4):67–94, July 1996.

88 C.R. Johnson, P. Schniter, T.J. Endres, J.D. Behm, D.R. Brown, and R.A. Casas. Blind equalization using the constant modulus criterion: A review. *Proceedings of the IEEE*, 86(10):1927–1950, October 1998.

89 S. Chen, A. Wolfgang, and L. Hanzo. Constant modulus algorithm aided soft decision directed scheme for blind space-time equalisation of SIMO channels. *Signal Processing*, 87:2587–2599, November 2007.

90 L. Zhang, W. Liu, and R.J. Langley. A class of constant modulus algorithms for uniform linear arrays with a conjugate symmetric constraint. *Signal Processing*, 90:2760–2765, September 2010.

91 F. Boccardi, R.W. Heath, A. Lozano, T.L. Marzetta, and P. Popovski. Five disruptive technology directions for 5G. *IEEE Communications Magazine*, 52(2):74–80, February 2014.

92 V. Venkateswaran and A.J. van der Veen. Analog beamforming in MIMO communications with phase shift networks and online channel estimation. *IEEE Transactions on Signal Processing*, 58(8):4131–4143, August 2010.

93 O. Oliaei. A two-antenna low-IF beamforming MIMO receiver. In *IEEE Global Telecommunications Conference*, pages 3591–3595, November 2007.

94 C. Miller, W. Liu, and R.J. Langley. Reduced complexity MIMO receiver with real-valued beamforming. In *IEEE International Conference on Computer and Information Technology*, pages 31–36, October 2015.

95 S. Han, C.I. I, Z. Xu, and C. Rowell. Large-scale antenna systems with hybrid analog and digital beamforming for millimeter wave 5G. *IEEE Communications Magazine*, 53(1):186–194, January 2015.

96 A.F. Molisch, V.V. Ratnam, S. Han, Z. Li, S.L.H. Nguyen, L. Li, and K. Haneda. Hybrid beamforming for massive MIMO: A survey. *IEEE Communications Magazine*, 55(9):134–141, September 2017.

97 F. Sohrabi and W. Yu. Hybrid analog and digital beamforming for mmWave OFDM large-scale antenna arrays. *IEEE Journal on Selected Areas in Communications*, 35(7):1432–1443, July 2017.

98 J. Li and P. Stoica, editors. *Robust Adaptive Beamforming*. John Wiley & Sons, Hoboken, New Jersey, 2005.

99 H. Cox, R.M. Zeskind, and M.M. Owen. Robust adaptive beamforming. *IEEE Transactions on Acoustics, Speech, and Signal Processing*, ASSP-35(10):1365–1376, October 1987.

100 M. Dogan and J.M. Mendel. Cumulant-based blind optimum beamforming. *IEEE Transactions on Aerospace and Electronic Systems*, 30(3):722–741, July 1994.

101 S.A. Vorobyov, A.B. Gershman, and Z.Q. Luo. Robust adaptive beamforming using worst-case performance optimization: A solution to the signal mismatch problem. *IEEE Transactions on Signal Processing*, 51(2):313–324, February 2003.

102 S. Shahbazpanahi, A.B. Gershman, Z.Q. Luo, and K.M. Wong. Robust adaptive beamforming for general-rank signal models. *IEEE Transactions on Signal Processing*, 51:2257–2269, September 2003.

103 J. Li, P. Stoica, and Z. Wang. On robust capon beamforming and diagonal loading. *IEEE Transactions on Signal Processing*, 51(7):1702–1715, July 2003.

104 Y. Zhao, W. Liu, and R.J. Langley. Robust broadband beamforming based on frequency invariance constraints and worst-case performance optimization. In *Proceedings of the International Symposium on Communications, Control and Signal Processing*, Limassol, Cyprus, March 2010.

105 L. Zhang and W. Liu. Robust forward backward based beamformer for a general-rank signal model with real-valued implementation. *Signal Processing*, 92:163–169, January 2012.

106 Y. Zhao and W. Liu. Robust wideband beamforming with frequency response variation constraint subject to arbitrary norm-bounded error. *IEEE Transactions on Antennas and Propagation*, (5):2566–2571, May 2012.

107 J.F. Sturm. Using SeDuMi 1.02, a MATLAB toolbox for optimization over symmetric cones. *Optimization Methods and Software*, 11:625–653, August 1999.

108 J. Lofberg. YALMIP: a toolbox for modeling and optimization in MATLAB. In *Proceedings of the IEEE International Symposium on Computer Aided Control Systems Design*, pages 284–289, Taipei, Taiwan, September 2004.

3

Electronically Steerable Parasitic Array

3.1 Introduction

A parasitic array antenna consists of a driven element and several parasitic elements [1]. The driven antenna is the only active radiating element, and the RF energy is distributed from the driven element to parasitic elements by electromagnetic coupling. One of the well known parasitic arrays is the Yagi-Uda antenna, which uses parasitic elements to obtain high directivity. There is another type of parasitic array antenna called the electronically steerable parasitic array (ESPAR). This type of parasitic array antenna applies tunable components to parasitic antenna elements to obtain reconfigurable radiation patterns. Figure 3.1 shows a typical example of a seven-element ESPAR antenna.

As shown in Figure 3.1, the element located at the center of the ESPAR antenna is the only active element connected to the RF front end. This active element is surrounded by a number of parasitic elements. The parasitic elements are loaded by electronically tunable components such as sliding loads, PIN diodes, varactors, and MEMS switches. The tunable loads can be adaptively controlled by a DSP. When performing the beamforming, the reactance loads of the parasitic array antennas are adjusted to perform the weighted signal synthesis. Since RF energy is distributed from the driven element to parasitic elements by electromagnetic coupling, the ESPAR antenna does not require any phase shifters or feed network. Furthermore, the ESPAR has one single RF front end only. Thus, it is a low-cost solution to smart antennas compared with traditional adaptive array antennas. In addition to the advantage of low cost, ESPAR antennas are also smaller in size and have lower weights compared to traditional adaptive array antennas.

In this chapter, the theory and operation principle of EPSAR antennas are presented first. Then various techniques of designing ESPAR antennas, including folded monopole ESPAR, low-voltage controlled ESPAR, planar ESPAR, etc., are discussed with examples. Some recent developments in ESPAR antennas are also presented. In section 3.6 two case studies, including practical designs for a compact monopole ESPAR and a planar ESPAR, are given.

3.2 Theory and Operation Principle

In an ESPAR antenna, each parasitic element is loaded with a reactive load (e.g. a varactor) that is controlled by a DC voltage. In the simplest case, there is one drive element

Low-cost Smart Antennas, First Edition. Qi Luo, Steven (Shichang) Gao, Wei Liu, and Chao Gu.
© 2019 John Wiley & Sons Ltd. Published 2019 by John Wiley & Sons Ltd.

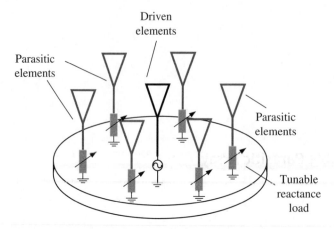

Figure 3.1 Configuration of the typical seven-element ESPAR antenna [2].

and n parasitic elements (e.g. monopoles) that are uniformly placed on a circular circumference of radius R, similar to the configuration shown in Figure 3.1. All the parasitic elements are loaded with tunable reactance loads, $(X_1, X_2, \dots, X_n)$. By tuning the reactance loads, the surface current on each parasitic element can be adjusted. Thus, the radiation pattern of the ESPAR antenna can be controlled. The distance (d) between the driven element and the parasitic elements should be kept around $\lambda/4$ in order to obtain the optimised mutual coupling and radiation performance of the array antenna [3]. Let the currents and voltages on the antenna elements be

$$I = [i_0, i_1, i_2, \dots i_n] \tag{3.1}$$
$$V = [v_0, v_1, v_2, \dots v_n] \tag{3.2}$$

where i and v represent the current and voltage on each antenna element, respectively. The current and voltage on the driven element are i_0 and v_0. Vectors I and V are associated with the mutual admittance matrix Y

$$I = Y \times V = \begin{bmatrix} y_{00} & y_{01} & \cdot & \cdot & y_{0n} \\ y_{10} & y_{11} & \cdot & \cdot & y_{1n} \\ \cdot & & \cdot & \cdot & \cdot \\ \cdot & & & \cdot & \cdot \\ y_{n1} & \cdot & \cdot & \cdot & y_{nn} \end{bmatrix} \times \begin{bmatrix} v_0 \\ v_1 \\ \cdot \\ \cdot \\ v_n \end{bmatrix} \tag{3.3}$$

where y_{ij} represents the mutual admittance between antenna elements i and j. The voltage vector V can also be calculated by using the equation

$$V = \begin{bmatrix} v_s \\ 0 \\ \cdot \\ \cdot \\ 0 \end{bmatrix} = \begin{bmatrix} jX_0 & 0 & \cdot & \cdot & 0 \\ 0 & jX_1 & \cdot & \cdot & 0 \\ \cdot & & \cdot & \cdot & \cdot \\ \cdot & & & \cdot & \cdot \\ 0 & 0 & \cdot & \cdot & jX_n \end{bmatrix} I \tag{3.4}$$

where v_s represents the transmitted voltage signal source with the amplitude and the phase from the driving RF port at the central element. Defining X and U_1 as

$$X = \begin{bmatrix} v_s \\ 0 \\ . \\ . \\ 0 \end{bmatrix} \tag{3.5}$$

$$U_1 = [1, 0, 0, .., 0]^T \tag{3.6}$$

the following expression can be obtained

$$I = v_s(Y^{-1} + X)^{-1}U_1 = v_s W \tag{3.7}$$

where W is the equivalent weight vector. The antenna generates a directional beam through tuning the reactance load on the parasitic elements. Signals transmitted from the central RF port excite the parasitic monopoles with substantial induced mutual currents on them. The far-field radiation pattern is formed by superposition of the radiation of all elements in the parasitic array. The far-field signal current in the azimuthal direction is represented as [3]

$$y(\theta)_{farfield} = I^T \alpha(\theta) = W^T v_s \alpha(\theta) \tag{3.8}$$

where $\alpha(\theta)$ is the steering vector

$$\alpha(\theta) = [1, e^{j\psi_1}, e^{j\psi_2}... e^{j\psi_n}] \tag{3.9}$$

where ψ_i is the phaser, which is [4]

$$\psi_i = \frac{2\pi d}{\lambda_0} \times \cos\left(\theta - 2\pi \frac{i-1}{n}\right) \tag{3.10}$$

where d is the distance between the driven element and the parasitic element.

In the ESPAR antenna, because each parasitic element is loaded with a reactive load, the reactances of the loads need to be optimised. The RF energy radiated from the driven antenna induces currents on the parasitic elements as a result of the mutual coupling. The current is transmitted through the transmission line, which is terminated with the reactive load. Assuming a varactor is used to load the parasitic elements, its impedance is Z_{var} and the impedance of the transmission line is Z_0, the reflection coefficient (Γ) of this loaded transmission line is

$$\Gamma = \frac{Z_{var} - Z_0}{Z_{var} + Z_0} \tag{3.11}$$

The corresponding reflection phase is

$$\Gamma_{phase} = \arctan \frac{Re(\Gamma)}{Im(\Gamma)} \tag{3.12}$$

where $Re(\Gamma)$ and $Im(\Gamma)$ are the real part and the imaginary part of the reflection coefficient, respectively. An ideal varactor is a reactive load; however, the commercial available packaged surface mount varactor is a combination of series and parallel circuits. Figure 3.2 shows the equivalent circuit of a varactor [5]. In this circuit model, R_s represents the series resistance, L_s represents the series inductance, and C_p is the parallel capacitance.

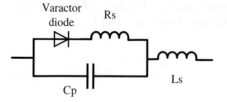

Varactor diode

Figure 3.2 The circuit model of the packaged varactor diode [5].

Table 3.1 Capacitance vs reverse voltage [6].

Voltage (V)	Typical capacitance (pF)
0	103.3
2	46.38
4	24.2
6	15.01
8	13.21
10	12.58

The total capacitance (C_T) is the parallel combination of the junction capacitance and the package capacitance. Its value is proportional to the applied voltage but the response is not linear. As an example, Table 3.1 shows the total capacitance of a varactor under different reverse voltages.

Therefore, to optimise the radiation performance of the ESPAR antenna, the reactance needs to be optimised. The objectives of the optimisation are to achieve higher total radiation efficiency and desired radiation patterns (beamforming). Moreover, the configuration of the parasitic array also needs to be optimised in order to obtain good impedance matching and appropriate mutual coupling between the driven and parasitic elements. Since there are many parameters involved in the optimisation process, using EM simulation tools with optimisation algorithms such as the genetic algorithm has been proposed and has demonstrated promising results [7].

Adaptive beamforming can be performed with a training sequence. The un-blinded algorithm enables the antenna to estimate the direction of the desired signal and form its null towards an interference signal. The normalised mean square error (NMSE) between a received signal and the reference signal is used as the objective function

$$Objective = 1 - |\rho|^2 \tag{3.13}$$

where ρ is the cross-correlation coefficient (CCC) between a received signal and the desired signal. The parameter ρ is defined as [3]

$$\rho = \frac{E[y(t) \cdot r(t)^*]}{\sqrt{E[y(t) \cdot y(t)^*]E[r(t) \cdot r(t)^*]}} \tag{3.14}$$

where $y(t)$ is the received signal and $r(t)$ is the desired signal. The desired signal $r(t)$ should be given to the baseband DSP controller in advance and operated as a reference.

The goal of the iteration is to minimise the NMSE. The steps of the optimisation are summarised below:

- The algorithm searches the best CCC value from the primary main patterns and determines the starting point for the following iteration.
- The algorithm iterates following the steepest gradient of CCC value. The control voltages supplied to the reactance loads are perturbed simultaneously [8]. Voltage components in the control voltage vector change at different speeds due to its sensitivity and contribution to the null forming at a particular position [9].
- When the NMSE between received signal and the reference signal is smaller than 0.001, the iteration threshold is met.

3.3 Low-cost Folded-monopole ESPAR

As presented in the previous section, the ESPAR antenna is a promising solution for low-cost smart antennas as there are no expensive phase shifters required and only one antenna element is connected with the RF front end, which avoids the need to use the feed network. Traditionally, an ESPAR antenna consists of one driven monopole and several parasitic monopoles with reactive loads [10, 11]. The volume of such a smart antenna is, however, too large for the application on portable devices. Thus, it is desirable to reduce the volume of the ESPAR antenna while maintaining its radiation performance.

To reduce the height of the monopole ESPAR antenna, one approach is to embed the ESPAR in a homogeneous dielectric material [12]. Figure 3.3 shows the concept of the dielectric loaded ESPAR antenna. By embedding a wire in a dielectric material, the physical size of an antenna can be reduced as a result of increased electrical size. This is a design technique that has been widely used in electrically small antenna designs [13, 14]. When the monopoles are embedded in a dielectric material, the effective wavelength can

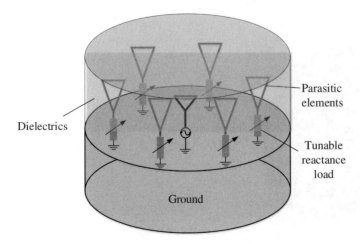

Figure 3.3 The concept of DE-ESPAR array antenna presented in [12].

be approximated as

$$\lambda_{eff} = \frac{\lambda_0}{\sqrt{\epsilon_r}} \tag{3.15}$$

where ϵ_r is the relative dielectric permittivity of the dielectric. Thus, the overall size of the ESPAR can be reduced by a factor of $\sqrt{\epsilon_r}$. As reported in [12], by embedding the ESPAR array in a cylindrical rod of the ceramic complex with a relative permittivity around 4.5, the volume of the array was reduced by 80% and the footprint was reduced by 50%.

Although using a dielectric to load the arrays can effectively reduce the volume of the ESPAR antenna, it increases the fabrication complexity and cost. Another simple but effective method to reduce the size of the ESPAR antenna is to use the folded monopole antennas as the parasitic elements and a top-disk loaded short monopole as the driven element [15, 16]. The folded monopoles are bent toward the driven element at the centre and provide the capacitive loads to the driven element. By taking advantage of the capacitive loading, the height of the ESPAR antennas can be reduced to less than $\lambda_0/10$ at the frequency of interest. Figure 3.4 shows the configuration of the folded monopole ESPAR (FM-ESPAR).

The top-disk loaded monopole at the centre is connected to a 50 Ω RF port. The six folded monopoles surrounding the driven element serve as reflectors. Each folded monopole is connected to the ground plane through a varactor. Pattern synthesis is determined by the surface currents on all radiating elements. The surface currents are controlled by varying the DC voltages supplied to the varactors. In order to increase the capacitive load, the distance between the driven element and the parasitic elements is kept small, which also reduces the radius of the ESPAR antenna. The sleeve ground plane is an optional component and the purpose of using the sleeve ground plane is to tune the direction of maximum radiation into the horizontal plane. Figure 3.5 shows an equivalent circuit representing the mutual coupling in the FM-ESPAR antenna. The six parallel elements that are connected to the secondary coil of the transformer represent the six parasitic folded monopoles.

Driven element

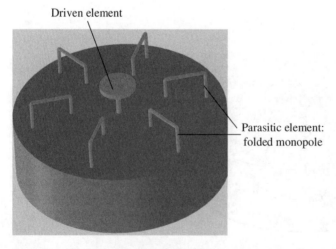

Parasitic element:
folded monopole

Figure 3.4 The configuration of the FM-ESPAR array antenna presented in [16].

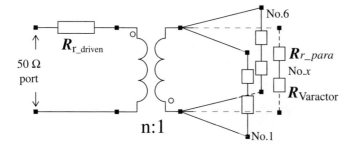

Figure 3.5 Mutual coupling equivalent circuit of a folded monopole ESPAR antenna. Courtesy of Dr Haitao Liu [17].

The total resistive load observed by the secondary coil can be calculated by

$$Z_L = \frac{1}{\sum_{i=1}^{6}(R_{i,var} + R_{i,para})} \tag{3.16}$$

where $R_{i,var}$ is the resistance of the varactor and $R_{i,para}$ is the parasitic radiation resistance of the folded monopole element i. If considering the coupling between the driven element and the parasitic elements as a wave propagating through a transmission line, the characteristic impedance Z_0 of the transmission line can be calculated as

$$Z_0 = \frac{1+\Gamma}{1-\Gamma}Z_L \tag{3.17}$$

where Γ is the reflection coefficient. Then, the distance between the driven element and the parasitic element (d) can be estimated by solving the equation because the distributed inductance and capacitance are functions of d

$$Z_0 = \sqrt{\frac{L}{C}} \tag{3.18}$$

$$L \approx \frac{\mu_0}{\pi} ln\left(2\frac{d}{D}\right) \tag{3.19}$$

$$C \approx \epsilon_0 \pi \frac{1}{ln\left(2\frac{d}{D}\right)} \tag{3.20}$$

where L and C are the equivalent distributed inductance and distributed capacitance, respectively. D is the wire diameter and d is the distance between the folded monopole and the driven monopole.

After roughly calculating the key parameters, such as the antenna height and the distance between the driven monopole and the folded monopoles, EM simulations are recommended to be performed to optimise the parasitic array antenna. Figure 3.6 illustrates the details of the antenna structure with the SubMiniature version A (SMA) connector and varactors. Table 3.2 shows the dimensions of the FM-ESPAR antenna (reported in [16]) calculated by using the equivalent circuit and the optimised values using EM simulators. As shown, the equivalent circuit gives a good estimation of the initial dimensions of the EM-ESPAR antenna.

As shown in Figure 3.6, because the FM-ESPAR antenna has a tightly coupled structure, the input impedance is sensitive to the values of the reactive loads. In order to

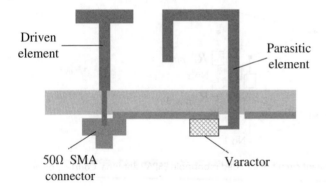

Driven element

Parasitic element

50Ω SMA connector

Varactor

Figure 3.6 The details of the antenna with SMA connector and varactor [17].

Table 3.2 Parameters of the FM-ESPAR presented in [16].

Parameters	Optimised by EM simulator	Calculated by equivalent circuit
Height of ESPAR antenna	11.10 mm (9.5+1.6)	11.00 mm
Total length of folded monopole	30.60 mm (29+1.6)	31.00 mm
Distance between the driven and parasitic elements	10.40 mm	9.98 mm

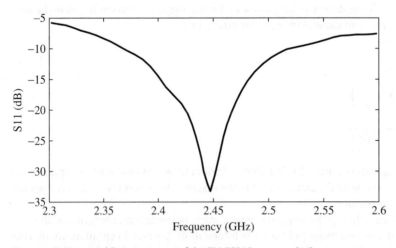

Figure 3.7 Measured S11 parameter of the FM-ESPAR antenna [16].

maintain a reasonable gain, the antenna is matched to the primary pattern mode, which is defined by supplying one varactor with a 22 V DC voltage and the other five varactors with 2 V DC voltages. The measured reflection coefficient is given in Figure 3.7.

Figure 3.8 shows the measured radiation patterns of the FM-ESPAR antenna operating in the primary pattern mode. The measurements were performed by applying a 22 V voltage on one varactor (one parasitic element) while the rest of the varactors were

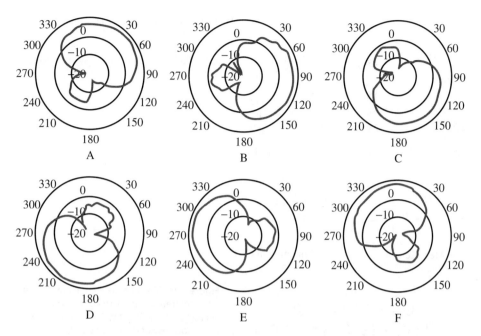

Figure 3.8 Measured radiation patterns of the FM-ESPAR antenna. Courtesy of Dr Haitao Liu [17].

Table 3.3 Beam direction and corresponding control voltages (in V).

Beam direction	V_1	V_2	V_3	V_4	V_5	V_6
30°	22	2	2	2	2	2
90°	2	22	2	2	2	2
150°	2	2	22	2	2	2
210°	2	2	2	22	2	2
270°	2	2	2	2	22	2
330°	2	2	2	2	2	22

supplied with 2 V voltages. As shown in this figure, under this configuration the main beam direction of the FM-ESPAR antenna is switched from 30° to 330° in 60° steps. The beam-switching is achieved by applying the 22 V voltage to the corresponding parasitic element, which leads to six different beams. During the beam switching, stable radiation patterns are observed. Table 3.3 summarises the beam directions of the antenna and its corresponding control voltages. The measured gain is 3.3 dBi and the front-to-back ratio (FBR) is 11.3 dB.

To increase the FBR, the adaptive backlobe cancellation method can be applied [18]. According to the calculation from [17], when applying the control voltage vector [23 V, 15 V, 3 V, 3V, 3 V, 15 V] to the varactors, the backlobe can be reduced and the gain can be increased. Figure 3.9 shows the measured and simulated radiation patterns of the FM-ESPAR antenna after applying the adaptive backlobe cancellation method. As can be seen, the FBR is increased to 20 dB and the maximum gain of the antenna is 4.0 dBi.

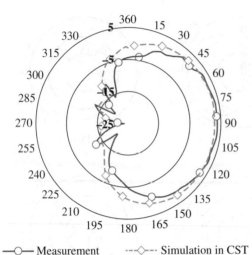

—○— Measurement   ---◇--- Simulation in CST

Figure 3.9 Measured and simulated radiation patterns of the FM-ESPAR antenna after applying the adaptive backlobe cancellation method. Courtesy of Dr Haitao Liu [17].

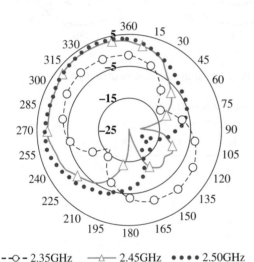

--○-- 2.35GHz   —△— 2.45GHz   •••• 2.50GHz

Figure 3.10 Measured radiation patterns at different frequencies with control voltage vectors 2.4 V, 2.4 V, 2.4 V, 2.4 V, 21.5 V, and 2.4 V. Courtesy of Dr Haitao Liu [17].

Because the radiation pattern of the ESPAR antenna is affected by the mutual coupling between the drive element and parasitic elements, which is related to the radiation resistance of the antenna element, for different frequencies, with the same condition of the applied voltages on the varactors, it cannot be guaranteed that the radiation patterns of the ESPAR antenna at different frequencies are optimised. As an example, Figure 3.10 shows the radiation patterns of the FM-ESPAR antenna at different frequencies under the control voltage vector [2.4 V, 2.4 V, 2.4 V, 2.4 V, 21.5 V, 2.4 V]. The best FBR and minimum backlobe is observed at 2.5 GHz. At 2.45 GHz and 2.5 GHz the gain is 4.0 dBi, whereas at 2.35 GHz the gain is −1 dBi, which implies that at this frequency the antenna has a very low efficiency. Therefore, to let the ESPAR antenna operate within the entire frequency band, the control weights of the FM-ESPAR antenna for different frequencies must be optimised individually. The optimisation can be done by changing the control voltage vector supplied to the varactors without adjusting any physical structure

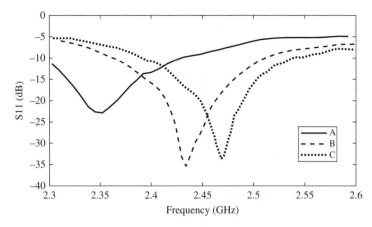

Figure 3.11 Measured S11 versus control voltage vectors [17]. The control voltages are A = [3 V, 3 V, 3 V, 3 V, 20 V, 3 V], A = [3 V, 3 V, 3 V, 3 V, 15 V, 3 V], and A = [3 V, 3 V, 3 V, 3 V, 24 V, 3 V].

of the antenna. In this way, the ESPAR antenna is able to operate over a wider frequency range as the resonance of the antenna becomes frequency reconfigurable. As an example, Figure 3.11 shows the measured reflection coefficient of the same FM-ESPAR with different sets of controlling voltage. During the measurement, only the high voltage applied to one of the parasitic elements is adjusted. As shown, the total operational bandwidth is from 2.3 to 2.55 GHz, which is three times as wide as the original bandwidth.

At each frequency there is an optimum control voltage vector, which gives an optimised radiation pattern. The optimised radiation patterns at 2.35 GHz, 2.45 GHz and 2.55 GHz, as well as their corresponding control voltage vectors, are given in Figure 3.12. These optimal radiation patterns are achieved by applying the control voltage vectors listed in Table 3.4. It is found that the antenna gain reduces as the frequency decreases, which can be regarded as the trade-off between optimising the bandwidth and improving the gain of the ESPAR antenna.

Figure 3.12 Optimized pattern at different frequencies. Courtesy of Dr Haitao Liu [17].

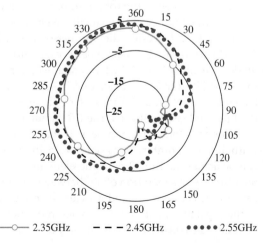

Table 3.4 Control voltages for optimised radiation patterns at different frequencies (in V).

Frequency (GHz)	V_1	V_2	V_3	V_4	V_5	V_6
2.35	3	3	3	3	3	15
2.40	3	3	3	3	3	17
2.45	3	3	3	3	3	20
2.55	9	9	2	2	9	24

3.4 ESPAR Antenna with Low Control Voltage

3.4.1 FM-ESPAR using PIN diodes

The FM-ESPAR antenna presented in Section 3.3 uses the varactors to tune the reactance of the parasitic antenna element. The advantage of using varactors is that they provide continuous capacitance variation within a certain range, which provides more degrees of freedom regarding beamforming. The disadvantage is that normally varactors require high DC control voltage (e.g. 0–22 V) to adjust their capacitances. The high DC control voltages required by varactors makes the FM-ESPAR antenna incompatible with most portable devices with a transistor–transistor logic (TTL) power supply. In order to design a low-profile ESPAR antenna that is compatible with TTL power supply, PIN diodes can be introduced to replace the varactors [19]. Unlike varactors which have continuous capacitance variation, the PIN diode has only two operational states: open (with reversely biased DC voltage) and short (with forward biased DC voltage). The advantage of using a PIN diode is that it requires much less power and its control voltage is compatible with the standard TTL control logic used by most low-power commercial wireless devices. The disadvantage of PIN diodes is that they only have two states, which constrains the beamforming capability.

The binary controlled beamforming concept [20, 21] can be applied to solve this problem. Since each parasitic element is loaded with a PIN diode, if an ESPAR antenna has N parasitic elements, the total number of different states is 2^N. Thus, for better beamforming, the number of the parasitic elements used in the ESPAR antenna is increased from 6 to 12. As a result, the number of available radiation patterns is increased from $2^6 = 64$ patterns to $2^{12} = 4096$ patterns. Figure 3.13 shows the configuration of the PIN diode controlled FM-ESPAR antenna, which has the same configuration compared to the FM-ESPAR antenna discussed in section 3.3 except that each parasitic element is loaded by a PIN diode for its surface current tuning. Since the insertion impedance of the PIN diode is small when the PIN diode is forward biased, the induced surface current along the folded monopole is stronger than when the PIN diode is reversely biased. With different combinations of forward and reversely biased PIN diodes, beamforming can be achieved. The 12 parasitic elements surrounding the centre are equally spaced on a circle with separation of $30°$ between adjacent elements. The parasitic elements are bent towards the centre and provide the capacitive load to the driven element. Thanks to the capacitive load from the folded monopoles, the height of the driven element can be reduced to $0.16\lambda_0$ at the frequency of interest [22].

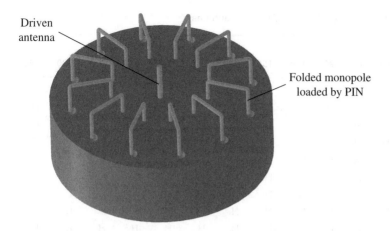

Driven antenna

Folded monopole loaded by PIN

Figure 3.13 Configuration of the PIN diode controlled FM-ESPAR antenna [22].

The beamforming of the antenna involves the synthesis of surface currents on the parasitic elements and the driven element. Because the amplitude of the induced current on the parasitic element when the PIN diode is forward biased is much larger than when the PIN diode is reversely biased, the parasitic elements with forward biased in-series PIN diodes dominate control of the antenna beamforming. The current in the azimuth direction is

$$Y(\varphi) = \vec{I}^T \cdot \alpha(\varphi) \tag{3.21}$$

where $\vec{I}^T$ represents the surface currents on the radiators and $\alpha(\varphi)$ is the steering vector

$$\alpha(\varphi) = \begin{bmatrix} 1 \\ e^{j\frac{d+L_B}{\lambda}2\pi cos(\varphi)} \\ e^{j\frac{d+L_B}{\lambda}2\pi cos(\varphi - \pi/6)} \\ . \\ . \\ . \\ e^{j\frac{d+L_B}{\lambda}2\pi cos(\varphi - 11\pi/6)} \end{bmatrix} \tag{3.22}$$

where L_B is the length of the bent part of the folded monopole.

When interference signals are considered, beamforming algorithms can be employed to choose a radiation pattern with the best SNR out of all the available radiation patterns. Each parasitic element has an individual control voltage. A 12-element DC voltage vector is represented by H and L, which correspond to 3 V and 0 V DC voltages, respectively. For example, a voltage vector [L, L, L, L, L, L, L, H, H, H, H, H] corresponds to the PIN diodes on folded monopoles 1 to 7 being reverse-biased and the PIN diodes on folded monopoles 8 to 12 being forward-biased. By right shifting the control voltage vector, the antenna's radiation pattern can be swept in the horizontal plane in steps of 30° within 360°. The relationship between the defined angle of the beams and the value of the DC voltage vectors is given in Table 3.5.

A beamforming example for one interference signal is given in Figure 3.14. In this example it is assumed that the desired signal comes from 90° and there is one interference signal from 130°. By configuring the DC control voltage vector as [L, H, H, H, L,

Table 3.5 Control voltage for the PIN loaded FM-ESPAR antenna.

Beam direction/PIN no.	30°	60°	90°	120°	150°	180°	210°	240°	270°	300°	330°	360°
1	H	H	L	L	L	L	L	L	L	H	H	H
2	H	H	H	L	L	L	L	L	L	L	H	H
3	H	H	H	H	L	L	L	L	L	L	L	H
4	H	H	H	H	H	L	L	L	L	L	L	L
5	L	H	H	H	H	H	L	L	L	L	L	L
6	L	L	H	H	H	H	H	L	L	L	L	L
7	L	L	L	H	H	H	H	H	L	L	L	L
8	L	L	L	L	H	H	H	H	H	L	L	L
9	L	L	L	L	L	H	H	H	H	H	L	L
10	L	L	L	L	L	L	H	H	H	H	H	L
11	L	L	L	L	L	L	L	H	H	H	H	H
12	H	L	L	L	L	L	L	L	H	H	H	H

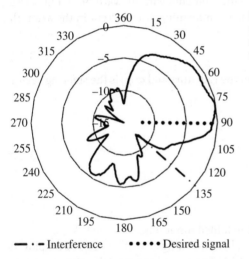

— · — Interference   • • • • • Desired signal

Figure 3.14 Measured normalised radiation pattern when the desired signal comes from 90° and the interference signal comes from 130°. Courtesy of Dr Haitao Liu [17].

L, H, H, L, L, L, L], a null is produced at the interference angle while the desired signal direction is within the 3-dB beamwidth of the antenna.

3.4.2 Link Quality Test

The smart antenna increases the channel capacity by providing spatial diversity and improves the link budget margin. To obtain a real-life measurement result, the PIN loaded FM-ESPAR antenna is tested in the wireless communication environment. The facts revealed by this test are two fold. First, the FM-ESPAR significantly increases the performance of a wireless system in both a multi-path environment and a large noise environment. Second, the antenna can be easily controlled with a simple 8-bit micro control unit (MCU), which is compatible with the low-cost system. The wireless transceiver chosen for this experiment is provided by ATMEL [23], following the IEEE 802.15.4 Standard [24], which is one of the most popular solutions for a short-range wireless communication system. An 8-bit MCU was used to control the transceiver as well as the FM-ESPAR using its 12 general purpose input/output pins. The test environment is full of instruments and workbenches with metal surfaces, which have a heavy multi-path effect on the experiment. It is difficult for a wireless system using such fading channels to establish a stable and fast link with the omnidirectional antenna. In the experiment, the system is configured to work in saturation mode, which means there is always a packet ready in the queue to be sent. The system throughput is therefore only affected by the channel quality, which enables a performance comparison of the PIN loaded FM-ESPAR antenna and omnidirectional antenna in the fading channel.

The system is configured to use the omnidirectional mode of the antenna first to show the normal wireless performance in such an environment. The receiver records the number of packets received for every single second, which is processed as the throughput. The system then uses the directional mode of the antenna to investigate the propagation environment around the antenna. The antenna is controlled to sweep in the horizontal plane with a speed of 30° every 5 seconds. It takes 60 seconds to finish the full 360° scan and this is repeated three times. The averaged link quality indicator (LQI) value is recorded by the MCU for each step and is given in Figure 3.15. As shown in this figure, beam no. 6 shows the best LQI, therefore the system employs the antenna with directional mode no. 6 to measure the throughput and the corresponding value is plotted in Figure 3.16. As shown, when the FM-ESPAR antenna is used, the achieved deliverable throughput is around 96k Kbps, which is much higher than the corresponding value using omnidirectional mode. With a simple algorithm, the system is able to lock the optimised angle to deliver the highest and most reliable throughput.

The antenna's interference rejecting ability was tested in the anechoic chamber with one interference scenario. The interference signal was an additive white Gaussian noise (AWGN) with 2 MHz bandwidth. The antenna was mounted at the centre of the anechoic chamber. The angle between the interference signal and the desired transmitted signal (Tx) was 80°. First, the antenna was configured to operate on its

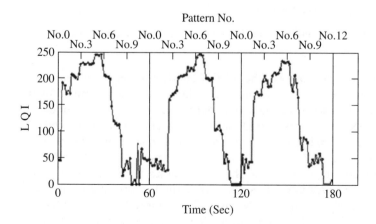

Figure 3.15 The LQI result. Courtesy of Dr Haitao Liu [17].

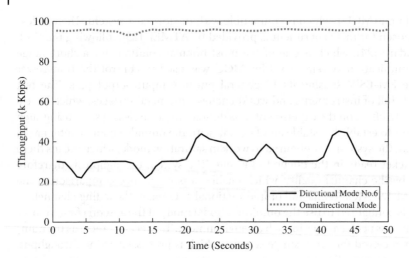

Figure 3.16 The throughput of the system, comparison between directional and omni-directional modes.

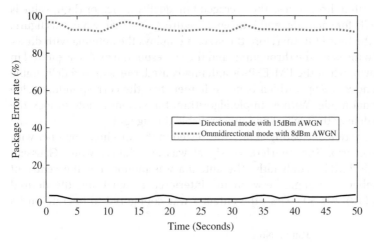

Figure 3.17 Package error rate compared between directional mode and omni-directional mode with one interference scenario.

omnidirectional mode and the power of the AWGN interference signal was increased until the communication between the Rx node and the Tx node was jammed by the AWGN. Second, the antenna was configured to operate in its directional mode and the power of the AWGN was increased until the communication was interrupted. The received package error rate was used to evaluate the link quality and the recorded values are plotted in Figure 3.17 for the omnidirectional mode with 8 dBm AWGN and the directional mode with 15 dBm AWGN. The omnidirectional mode was nearly jammed when the power of AWGN was 8 dBm, while the link was still reliable with directional mode when the power of AWGN was 15 dBm. This test shows that the PIN loaded FM-ESPAR provides a reliable link for a wireless communication system in a large noise environment.

3.5   Planar ESPAR Antennas

Traditional ESPAR antennas are based on wire antenna elements, such as monopoles. In many applications it is highly desirable to have the smart antenna in planar form, which is better for system integration and packaging. Another advantage of using a planar configuration is to realise a circularly polarised ESPAR design because it is difficult to realise circular polarisation on a wire antenna.

Several types of planar parasitic arrays were designed based on the Yagi-Uda concept. A linearly polarised three-element microstrip parasitic array antenna was presented in [25]. This antenna consists of one driven strip and two parallel parasitic strips, which are printed on a grounded dielectric substrate. Figure 3.18 shows the configuration of this planar parasitic array. The driven element is fed by an SMA probe and its length is approximately half of a guided wavelength at its resonance. The distance between the driven and parasitic elements is a quarter-wavelength in free space, which is the value used to design the monopole ESPAR antenna and Yagi–Uda antenna. Beam switching can be achieved by reconfiguring the electrical length of the parasitic strips by changing the states of the RF switches. Switching on the two RF switches that are inserted into the gaps of the parasitic element, the strips work as a reflector. On the other hand, by switching off the two RF switches the parasitic strips work as a director. Figure 3.19 shows the equivalent Yagi–Uda antenna configuration when the parasitic elements work in either the reflector or director modes. In these two modes, the antenna's maximum radiation direction can be shifted towards a maximum angle of ±35° in the H-plane over an overlapped bandwidth from 3.67 to 3.865 GHz, corresponding to a fractional bandwidth of 5.2% . When all the switches are turned off, the antenna has a broadside radiation pattern with good impedance matching.

Another work based on the microstrip Yagi–Uda array is a circularly polarised crossed patch array [26]. This antenna makes use of the microstrip Yagi antenna to tilt the beam of the parasitic array from the broadside. The design rule for the patch-type Yagi array involves the optimisation of the dimension ratio between the director patch and the driven element patch, as well as the distance between the patches. Figure 3.20 shows the structure of the antenna. It consists of two identical orthogonal linear patch Yagi arrays and the driven element is positioned in the centre. As an essential technique used in this work, the four slots that are incorporated with the MEMS switches can be reconfigured to change the resonance frequency of the parasitic patches. The reflective and directive effect of the reconfigurable parasitic patches is validated for linear polarisation

Figure 3.18 The concept of the three-strip planar parasitic array antenna [25].

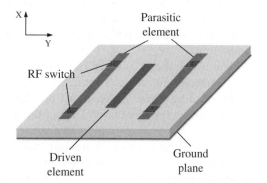

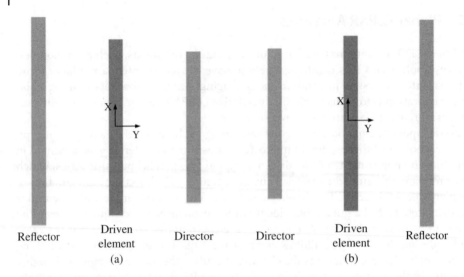

| Reflector | Driven element | Director | Director | Driven element | Reflector |

(a)   (b)

Figure 3.19 The equivalent Yagi–Uda antenna configuration when the parasitic array works in different modes: (a) the H-plane radiation pattern tilts toward to the positive *y* axis and (b) the H-plane radiation pattern tilts toward to the negative *y* axis.

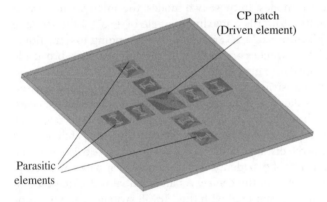

CP patch
(Driven element)

Parasitic
elements

Figure 3.20 The configuration of the microstrip Yagi–Uda parasitic array antenna presented in [26].

first. When the two feed points of the driven patch are simultaneously excited, circularly polarised radiation is obtained. By activating different operating modes, four switched beams with a maximum tilt angle of 30° can be generated, covering the whole horizontal plane.

Figure 3.21 shows another two typical configurations of the planar ESPAR using microstrip antennas. Both configurations consist of a driven element at the centre surrounded by parasitic elements. Both the active element and the parasitic elements are printed on the same layer. Each of the parasitic elements is loaded with a variable reactance.

The type of antenna element is not only limited to microstrips, it can also be slot antennas, dielectric resonators, etc. For example, an ESPAR antenna using parasitic slot array antennas was presented in [27]. In this design, the array consists of three linearly spaced parasitic slot antennas, one driven slot, and two parasitic slots, as demonstrated

Figure 3.21 Two configurations of the planar ESPAR antenna: (a) four parasitic elements and (b) six parasitic elements.

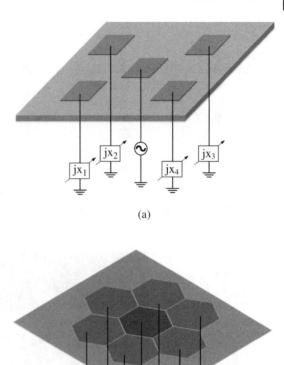

(a)

(b)

in Figure 3.22. The length of the slot is a half-wavelength at the frequency of interest and it is electromagnetically fed by a 50 Ω microstrip. Each of the two parasitic slots is loaded with a variable reactance, which is placed at the end of the microstrip feed line.

As presented in the last section, the mutual coupling between the active element and the parasitic elements is critical for the operation of the ESPAR antenna. In the case of the planar ESPAR, the mutual coupling is weaker than the monopole based ESPAR antenna because of its planar configuration. This means that the mutual impedances shown in Equation 3.3, which is re-written below, have smaller values than monopole based ESPAR antennas. Thus, the distance between the drive and parasitic elements need to be optimised and the optimum value may be smaller than the quarter-wavelength

$$I = Y \times V = \begin{bmatrix} y_{00} & y_{01} & \cdots & y_{0n} \\ y_{10} & y_{11} & \cdots & y_{1n} \\ \cdot & \cdot & \cdots & \cdot \\ \cdot & \cdot & \cdots & \cdot \\ y_{n1} & \cdot & \cdots & y_{nn} \end{bmatrix} \times \begin{bmatrix} v_0 \\ v_1 \\ \cdot \\ \cdot \\ v_n \end{bmatrix} \quad (3.23)$$

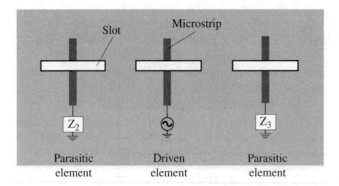

Figure 3.22 Configuration of the ESPAR using a slot antenna [27].

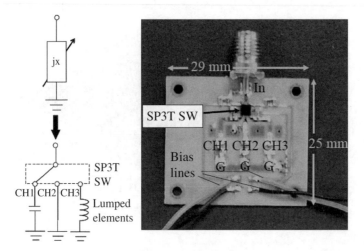

Figure 3.23 The circuit of the reactance load of parasitic elements and a photograph of the circuit [28]. Reproduced with permission of ©EurAPP.

A five-element circularly polarised planar ESAPR was reported in [28]. This array has the same configuration as that shown in Figure 3.21a. Through an experimental study of the reactive load, it is found that good beam-switching can be achieved by loading the parasitic element with three different states, namely short, inductor, and capacitor, as shown in Figure 3.23. As shown in this figure, this circuit uses an SP3T switch to adjust the load between the three states $(-jx, j0, +jx)$. When the parasitic element is shorted, it does not contribute to the radiation and acts almost as transparent to the driven element. When the parasitic element is loaded by either an inductive or a capacitive load, it acts as the director with phase delays of different sign.

By setting the reactance with a proper value, the planar ESPAR antenna reported in [28] is able to tilt the main beam in the elevation plane around the azimuth of every $45°$ with good circular polarisation (AR < 3 dB). Figure 3.24 shows the measured radiation pattern and axial ratio of this five-element circularly polarised ESPAR at the $\phi = 0°$ plane when the reactance value is set as $(-j60\Omega, -j60\Omega, +j60\Omega, +j60\Omega)$. The main beam of the

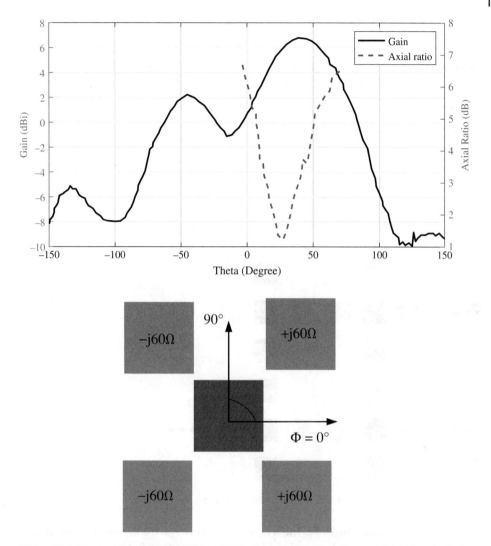

Figure 3.24 The measured radiation pattern and axial ratio of the five-element circularly polarised ESPAR at the $\phi = 0°$ plane [28]. The corresponding reactance values are shown.

array is steered to 38° with a gain of around 7 dBic, while the axial ratio at the main beam direction is around 2 dB.

The planar ESPAR antenna presented in [28] can achieve eight different beams with axial ratio less than 3.5 dB. Table 3.6 summarises the measured beam-switching angle, gain and corresponding axial ratio of this five-element ESPAR antenna.

One of the advantages of planar ESPAR antennas is that they can be easily extended to the design of array antenna of large size. This can be realised by using the ESPAR antenna as the subarray, as demonstrated in Figure 3.25.

To steer the beam to the desired direction (θ, ϕ), the reactive loads of the parasitic elements in each of the subarrays first need to be configured to steer the beam to this angle. Then the driven element of each subarray needs to be excited with the required

Table 3.6 Measured beam-switching performance of the five-element ESPAR antenna.

Beam direction (ϕ)	Maximum gain angle (θ)	Gain (dBic)	Axial ratio (dB)
−135°	31°	5	1.2
−90°	36°	6.5	1
−45°	29°	5.5	2.8
0°	38°	7	3
45°	28°	5	3.5
90°	33°	6.5	2
135°	28°	6	2.5
180°	36°	7	2.3

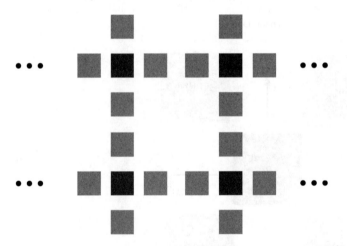

Figure 3.25 Use of an ESPAR array to form an antenna array of larger size [29].

phase to direct the beam to (θ, ϕ). The relative phase between the driven element and the parasitic element in each subarray should be kept the same as what has been configured in the first step. As a result, the phase shifters have to be incorporated in the driven elements. Moreover, attenuators can also be included to control the amplitude distribution of the array antenna at the subarray level, which can be used to optimise the radiation pattern of the array, such as reducing the sidelobe levels. Therefore, using this approach, compared to traditional phased arrays, the number of phase shifters or RF chains can be reduced by a factor of N, where N is the number of parasitic elements. This greatly reduces the cost of the antenna system. Figure 3.26 demonstrates the concept of using the planar ESPAR antenna to form an array antenna of larger size.

One drawback of using a subarray to form a large size array is the appearance of grating lobes. This is because the distance between the antenna elements is large at some planes (e.g. the diagonal plane). The same problem also occurs in phased array antenna design when the subarray technique is used. For the phased array, the grating lobes can be eliminated by applying amplitude tapering thanks to its sophisticated beam forming

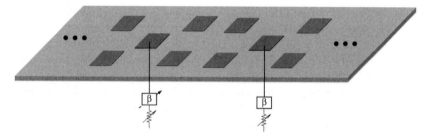

Figure 3.26 Configuration of incorporating a phase shifter and attenuator in the driven elements of the ESPAR array to form an antenna array of larger size.

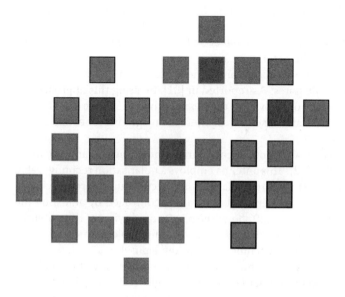

Figure 3.27 Use of an ESPAR array with an interleaved configuration to form an antenna array of larger size [29].

network [30]. However, for ESPAR antenna, this solution is not feasible because it does not have any feed network. One approach that is applicable to the ESPAR antenna to eliminate the grating lobes is to use the concept of an interleaved configuration [29], as shown in Figure 3.27.

The mutual couplings between the driven element and the parasitic elements are important for the operation of ESPAR antennas. As shown section 3.3, Figure 3.11, at different beam steering angles the impedance matching of the antenna varies. This is because the reflection coefficient is related to the mutual admittance between the driven antenna and the parasitic elements

$$
\begin{bmatrix}
y_{00} & y_{01} & \cdots & y_{0n} \\
y_{10} & y_{11} & \cdots & y_{1n} \\
\cdot & \cdot & \cdots & \cdot \\
\cdot & \cdot & \cdots & \cdot \\
y_{n1} & \cdot & \cdots & y_{nn}
\end{bmatrix}
\tag{3.24}
$$

where y_{ij} represents the mutual admittance between antenna element i and antenna element j. Although by optimising the applied voltages on each parasitic element (to adjust the value of the reactance) the bandwidth of the antenna can be tuned to cover the frequency band of interest at different beam steering angles, it is noticed that the resonance of the antenna actually shifts to other frequencies. This can be compensated by controlling the mutual couplings using lumped elements. Thanks to the planar configuration, the mutual couplings can be controlled by introducing surface mount varactors between the driven element and the parasitic elements [31]. The advantage of this method is that by varying the mutual couplings the resonances of the antenna can be kept at the same frequency on different operation modes of the ESPAR antenna. Figure 3.28 presents this concept. As shown, varactors are placed between the driven and parasitic patches. Compared to the monopole based ESPAR antenna, the planar ESPAR antenna can easily integrate the surface mount capacitors between the driven and parasitic elements. The capacitances of the varactors, which have an effect on the mutual coupling between the patches and the resonances of the antenna, can be controlled as an additional degree of design freedom to compensate for variation in the mutual coupling when the planar ESPAR operates at different states. As reported in [31], by using this approach the planar ESPAR antenna can electronically scan the beam within ±15 while maintaining good impedance matching and stable resonances. There are no grating lobes and a peak gain of 7.4 dBi was obtained.

A microstrip ESPAR antenna with both reconfigurable polarisations and reconfigurable patterns was reported in [32]. This antenna employs a polarisation reconfigurable microstrip patch as the driven element. The polarisation is reconfigured through changing the polarisation of the driven element and its beam is controlled by adjusting the loads of the parasitic elements. Figure 3.29 shows the plan and side view of this antenna. As shown, the driven patch is surrounded by eight reconfigurable parasitic dipoles, which are used to realise the beam-switching. Four of the eight dipoles are responsible for one polarisation. The reason for introducing the two-layer parasitic dipoles is to generate more beams and achieve a larger beam tilt angle. For example, when the driven patch is horizontally polarised, dipoles H1, H2, H3, and H4, which are placed along the horizontal direction, are excited. As a result, the beam control mainly depends

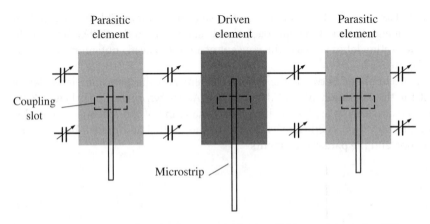

Figure 3.28 The concept of an ESPAR antenna with controllable mutual coupling [31].

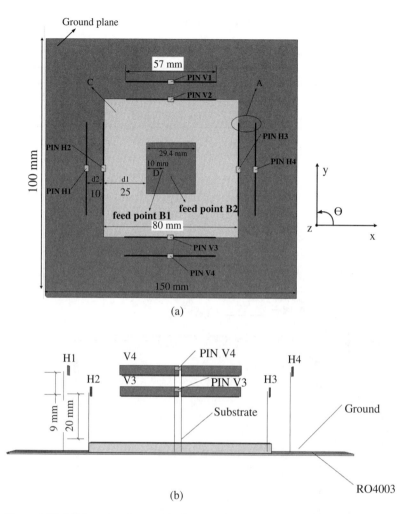

Figure 3.29 (a) Plan view. A, parasitic elements; B1, B2, microstrip antenna feed points; C, substrate; d1, distance from the dipole to patch; d2, distance between the two parasitic elements . (b) Side view. H1, H4, V4, distance from the lower parasitic element to the upper parasitic element; H2, H3, V3, distance from the patch to the lower parasitic element [32].

on the states of the four PIN diodes integrated on these four dipoles. Table 3.7 shows the detailed PIN diode switch configurations under which the three beams are obtained.

In order to minimise the spurious radiation from the feed lines and the impact of the feed network, the microstrip antenna is fed by two probes. Thus, the ground plane of the patch antenna can act as a good shield to isolate the unwanted radiation from the DC bias lines. The key to realising the polarisation reconfiguration is to design a two-way DC feed line with RF chokes. The detailed feed network is depicted in Figure 3.30. Three capacitors are used to isolate the DC currents induced by the bias lines. Moreover, quarter-wavelength high impedance lines along with a radial stub are designed to create an AC short-circuited point. To present a high impedance at the design frequency while allowing a DC biasing for a PIN diode (BAR 64-02V), inductors should be

Table 3.7 Beam and PIN diode switch configurations (orthogonal polarisations) [32].

	PIN H1/H2	PIN H3/H4	PIN V1/V2/V3/V4
Beam 1 ($\phi = 0°$)	Reverse biased	Forward biased	Reverse biased
Beam 2 ($\phi = 180°$)	Forward biased	Reverse biased	Reverse biased
Beam 3 ($\phi = 0°, 180°$)	Forward biased	Forward biased	Reverse biased

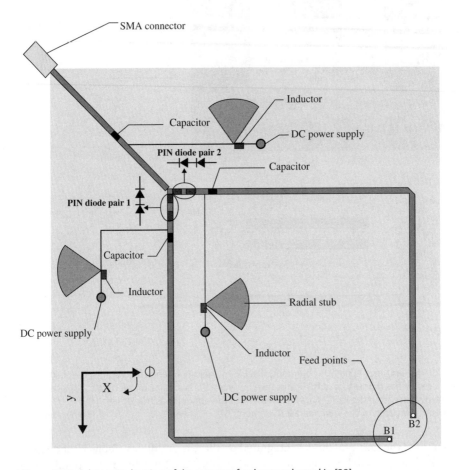

Figure 3.30 Schematic drawing of the antenna feed network used in [32].

carefully selected as the RF chokes. With a self-resonant frequency higher than 2.4 GHz, a 56nH inductor is chosen for this application. After incorporating these components, the two microstrip lines are connected to the probes which feed the patch. To realise high isolation between the two ports, two series PIN diodes are used for each feed line. By independently controlling the PIN diode pairs 1 and 2 in Figure 3.29, the feed points B1 and B2 are selected accordingly and thus the polarisation of the antenna can

Table 3.8 Polarisation direction and PIN diode switch configurations [32].

	PIN pair 1	PIN pair 2
Horizontal polarisation (along x axis)	Forward biased	Reverse biased
Vertical polarisation (along y axis)	Reverse biased	Forward biased
Diagonal polarisation	Forward biased	Forward biased

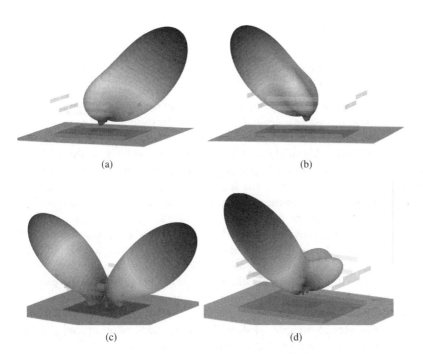

(a)   (b)

(c)   (d)

Figure 3.31 The 3D radiation patterns of the polarisation reconfigurable ESPAR antenna: (a) horizontal polarised, beam steered to ($\theta = 45°, \phi = 0°$), (b) horizontal polarised, beam steered to ($\theta = -45°, \phi = 180°$), (c) horizontal polarised, beam steered to $\phi = 0°$ and $\phi = 180°$, and (d) diagonal polarised, beam steered to ($\theta = -45°, \phi = 135°$). For demonstration purposes the radiation patterns are plotted using a linear scale.

be switched. Table 3.8 shows the PIN diode switch configurations under which the three polarisations are obtained.

Figure 3.31 shows the 3D radiation patterns of this polarisation reconfigurable ESPAR antenna. Because of the symmetrical configuration, only vertical and diagonal polarisations are shown in this figure. When the antenna is horizontally polarised, as shown in Figure 3.31a–c, the ESPAR can be configured to operate at two single beam modes: ($\theta = 45°, \phi = 0°$) and ($\theta = -45°, \phi = 180°$). Meanwhile, it also has a dual-beam mode that has two simultaneous beams pointing at ($\theta = 45°, \phi = 0°$) and ($\theta = -45°, \phi = 180°$). When the driven patch is diagonally polarised, the beam can be steered to either ($\theta = -45°, \phi = 135°$) or ($\theta = 45°, \phi = 45°$). In Figure 3.31d, only one mode is presented due to the symmetrical configuration.

3.6 Case Studies

In this section two case studies are presented. The first case study demonstrates the design and optimisation of the monopole ESPAR and the use of an additional dielectric layer to reduce the volume of the array. The second case study is a planar ESPAR with a thin thickness. This antenna has a much lower profile and the beam-switching is achieved by controlling the states of the PIN diodes.

3.6.1 ESPAR using Monopole

This case study presents a classic ESPAR design which uses monopoles as the antenna elements. Figure 3.32 shows the configuration of the monopole ESPAR with seven elements. As discussed in this chapter, the key parameters for the monopole ESPAR are:

- The length of the monopole. This determines the resonant frequency of the monopole. Since the monopole is placed on a ground skirt, its height should be approximately a quarter-wavelength at the frequency of interest.
- The distance between the driven element and the parasitic elements. This distance is critical in determining the mutual coupling between the driven element and the parasitic elements. It can also affect the input impedance of the driven element. The typical distance for the optimum radiation performance is approximately a quarter-wavelength at the central frequency.
- The reactive loads of the parasitic elements. The reactive load is the most important parameter to determine the radiation pattern of the ESPAR. It also has an effect on the reflection coefficient of the driven element. Due to the complex electromagnetic coupling between the antenna elements, the values of the reactive loads are difficult to calculate and EM simulation with optimisation algorithms is normally required to obtain the optimum value of these loads.
- The height and diameter of the ground skirt. A ground plane of optimised size can help improve the radiation pattern of the antenna.

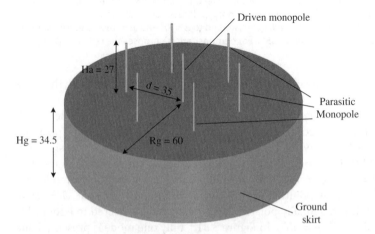

Figure 3.32 The configuration of the seven-element monopole ESPAR. The dimensions are in millimetres.

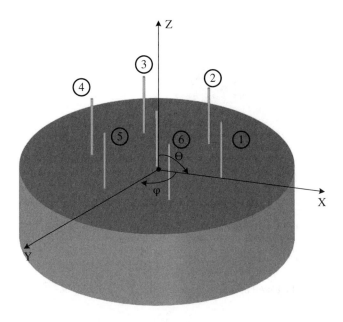

Figure 3.33 The coordinates of the seven-element monopole ESPAR.

In this design example, the operating frequency is chosen to be 2.45 GHz and the objective is to steer the beam of the antenna to $(\theta = 90°, \phi = 0°)$. The initial antenna parameters are taken from [7] as the design reference. The driven and parasitic elements have the equal length, which is 27 mm $(0.22\lambda_{2.45GHz})$. The distance between the driven and the parasitic element is 35 mm $(0.28\lambda_{2.45GHz})$. The diameter of the ground skirt is 60 mm and its height is 34.5 mm. These parameters are detailed in Figure 3.32 and Figure 3.33 shows the corresponding coordinates.

To steer the beam of the ESPAR to the desired angle, the values of the reactive loads need to be optimised. The genetic algorithm was used in [7] to optimise the values of the reactive loads. The cost function was defined to search the maximum gain of the ESPAR at the required beam angle and the stopping criterion of the optimiser was the maximum generation limit. Table 3.9 lists the value of the optimised values of the reactive loads and Figure 3.34 shows the corresponding 3D radiation pattern. The maximum directivity of the antenna is about 8.6 dB with a 3 dB beamwidth of 70°.

Figure 3.35 shows the simulated realised gain when the distance between the driven and parasitic elements varies. It can be seen that the value of d is critical in deciding the realised gain of the antenna. If the distance between the driven and parasitic elements is not optimised, the ESPAR antenna would suffer low efficiency. For example, if d is chosen to be 25 mm, the realised gain of the antenna is only 5,dB which is 2.3 dB less than the optimised design.

Table 3.9 Optimised reactive loads of the seven-element monopole ESPAR [7].

Antenna number	1	2	3	4	5	6
Equivalent value of the lumped element	4.7 pF	1 pF	2.6 nH	1.1 nH	2.6 nH	1 pF

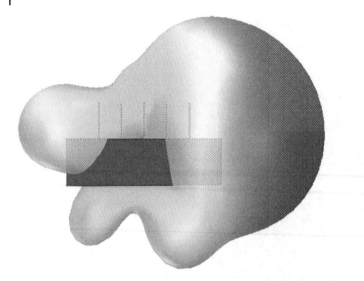

Figure 3.34 The 3D radiation pattern of the ESPAR using the reactive loads listed in Table 3.9.

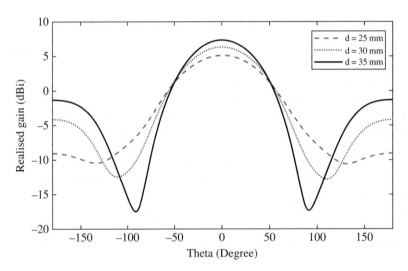

Figure 3.35 The simulated radiation pattern when the distance (*d*) between the driven and parasitic elements varies.

Parametric studies were performed to investigate the optimum height of the ground skirt and the simulation results are shown in Figure 3.36. As shown, although the height of the ground skirt can affect the radiation pattern of the ESPAR, its effect is not significant.

To reduce the size of the ESPAR, a dielectric layer is introduced and the antenna elements are embedded in the dielectric radome. In this example, the relative permittivity ε_r of the dielectric is chosen to be 3.55. With this configuration, the height of the monopole, the distance between the driven and parasitic elements, can be reduced by a factor of $\sqrt{\varepsilon_r}$. Figure 3.37 shows the configuration of the dielectric loaded ESPAR.

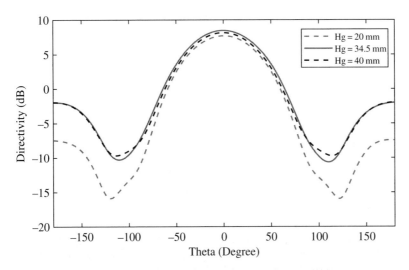

Figure 3.36 The simulated radiation pattern when the height of the ground skirt (H_g) varies.

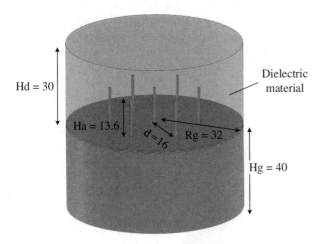

Figure 3.37 The configuration of the seven-element dielectric loaded monopole ESPAR. The dimensions are in millimetres.

For this design, the quasi-Newton method [33] is used to optimise the values of the reactive loads. The design objective is kept the same, to steer the main lobe of the ESPAR to ($\theta = 90°, \phi = 0°$). The fitness function is defined as the directivity at ($\theta = 90°, \phi = 0°$) higher than 7 dBi

$$D(\theta = 90°, \phi = 0°) \geqslant 7\text{dBi} \tag{3.25}$$

Meanwhile, because of the symmetrical configuration, the reactive loads should satisfy

$$X_2 = X_6 \tag{3.26}$$
$$X_3 = X_5 \tag{3.27}$$

Table 3.10 Optimised reactive loads of the seven-element monopole ESPAR embedded in the dielectric.

Antenna number	1	2	3	4	5	6
Equivalent value of the lumped element	10 pF	0.5 pF	4.5 nH	0.8 nH	4.5 nH	0.5 pF

where X_i is the reactance of parasitic element i. By setting up the optimiser with the above fitness function and conditions, the optimisation goal can be reached within 35 minutes (170 iterations) using a desktop PC with 8-core CPU and 32G RAM. Table 3.10 lists the values of the reactive loads and Figure 3.38 shows the optimised radiation pattern of the ESPAR antenna.

With the current configuration, which has six parasitic elements, the beam of the ESPAR can be switched in the entire azimuth plane at the 60° step, as shown in Figure 3.39. It can be seen that the entire azimuth plane is within the 3 dB beamwidth coverage of the antenna, and the beam overlap is about 1.5 dB.

To get better resolution, more parasitic elements can be introduced. As an example, Figure 3.40 shows the dielectric loaded ESPAR with 12 parasitic elements. The dimensions of the antenna elements are kept the same as the one presented in Figure 3.37. The quasi-Newton optimisation algorithm is performed to optimise the radiation pattern using the same fitness function and symmetrical conditions defined above. Figure 3.41 shows the optimised radiation pattern and Table 3.11 lists the value of each reactive load.

3.6.2 Planar Ultra-thin ESPAR

In this case study, a planar ultra-thin ESPAR [34] using a crossed dipole as the driven element is presented. The crossed dipole is chosen as the driven element because the

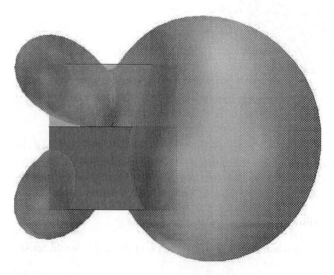

Figure 3.38 The optimised radiation pattern of the dielectric loaded ESPAR antenna with seven elements.

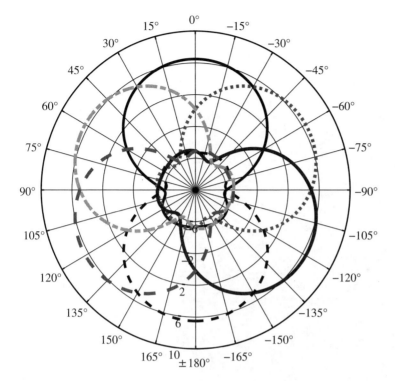

Figure 3.39 The beam-switching performance of the dielectric loaded seven-element monopole ESPAR.

Table 3.11 Optimised reactive loads of the 12-element monopole ESPAR embedded in the dielectric material.

Antenna number	1	2	3	4	5	6
Equivalent value of the lumped element	4.5 pF	5.6 pF	0.75 pF	4 pF	1 nH	4.2 nH
Antenna number	7	8	9	10	11	12
Equivalent value of the lumped element	1.3 nH	4.2 nH	1 nH	4 pF	0.75 pF	5.6 pF

E-field of orthogonally crossed dipoles in phase quadrature is linearly polarised at each specific observation point in the far-field in the azimuth plane. This characteristic can be utilised to design a beam-switching ESPAR by introducing parasitic elements in the same plane as the crossed dipole. This concept is illustrated in Figure 3.42. As shown, the crossed dipole is surrounded by four parasitic dipoles. Each dipole is loaded by a PIN diode. By turning on the four diodes sequentially, this antenna is expected to switch between four directional patterns.

The operation principle of this planar ultra-thin ESPAR is illustrated in Figure 3.43. A directional radiation pattern can be obtained if the driven element is backed by a reflector and guided by a director. This is equivalent to a crossed dipole driven Yagi–Uda array. Since the magnitude of the rotating E-field radiated by orthogonally crossed dipoles in

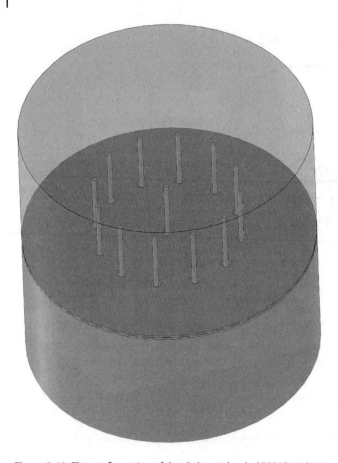

Figure 3.40 The configuration of the dielectric loaded ESPAR with 12 parasitic elements.

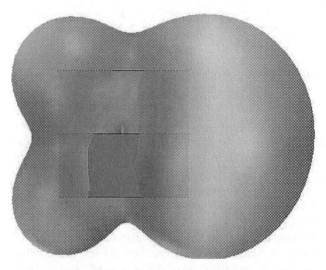

Figure 3.41 The optimised radiation pattern of the dielectric loaded ESPAR antenna with 12 parasitic elements.

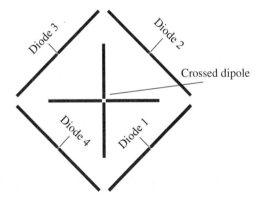

Figure 3.42 Concept of switched beams using orthogonally crossed dipoles in a phase quadrature driven ESPAR.

phase quadrature is non-zero at all azimuth angles, continuous beam switching can be achieved when the equivalent reflector and director are rotated by means of changing the states of PIN diodes.

Figure 3.44 shows the configuration of the planar ultra-thin ESPAR. The antenna is printed on both sides of a 0.813 mm thick Rogers RO4003C substrate with relative dielectric permittivity of 3.55 and dissipation factor of 0.0027. On the top layer, the fan-shaped arm 1 and arm 2 are connected by a 3/4 ring-shaped phase delay line, which is then connected to the inner pin of the coaxial cable. The 3/4 ring-shaped line introduces a 90° phase difference between

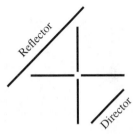

Figure 3.43 An orthogonally crossed dipole driven Yagi–Uda array.

the crossed dipoles, which ensures the radiated E field is relatively even in the azimuth plane. A four gap and four diodes loaded ring is placed outside the fan-shaped arms, which yields four parasitic elements. To control each PIN diode, two high impedance

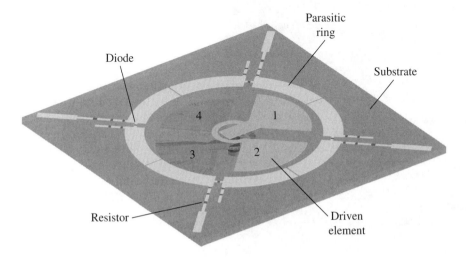

Figure 3.44 The configuration of the planar ultra-thin ESPAR.

microstrip lines are deployed perpendicularly to the ring. On the DC biasing lines for each diode there are four inductors to choke the RF signal and one resistor to limit the current through the PIN diode. On the bottom layer of the substrate, arms 3 and 4 are connected to each other through the phase delay line and are placed symmetric to arms 1 and 2. As the opposite fan-shaped arm pair forms a half-wavelength dipole, the electrical length of each fan-shaped arm should be around $\lambda_g/4$ at the design frequency. Both the central angle β and radius R_3 determine the actual electrical length of the arm. This antenna is designed to resonate at 2.3 GHz and the important parameters are labelled in Figure 3.45.

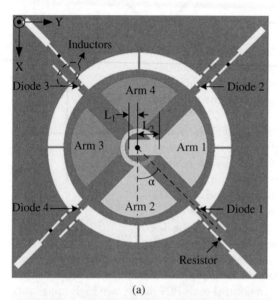

Figure 3.45 Antenna configuration and dimensions: (a) top view and (b) bottom view. The values of the parameters are $R_1 = 4, R_2 = 5.5, R_3 = 16, R_4 = 22$, $W_1 = 4.5, W_2 = 0.42, W_3 = 2.13, W_4 = 5$, $L_1 = 3$, and $W_2 = 6.25$.

(a)

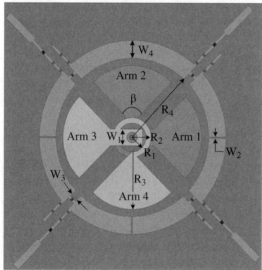

(b)

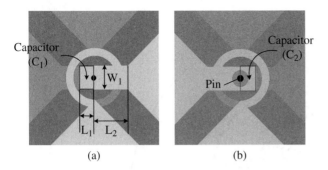

Figure 3.46 Configuration of overlapped rectangular patch: (a) top view and (b) bottom view.

The impedance matching of this antenna is obtained by introducing additional capacitance to the driven element. This is achieved by using a pair of rectangular patches that are printed on both sides of the substrate with an overlapped area of $W_1 \times L_1$, as shown in Figure 3.46. This overlapped structure of two patches results in a pair of parallel-plate capacitors and through varying the size of the overlapped patch the impedance matching of the driven element can be optimised. The simulated input impedance of the antenna for various values of L_1 is shown in Figure 3.47. As shown, the input impedance without

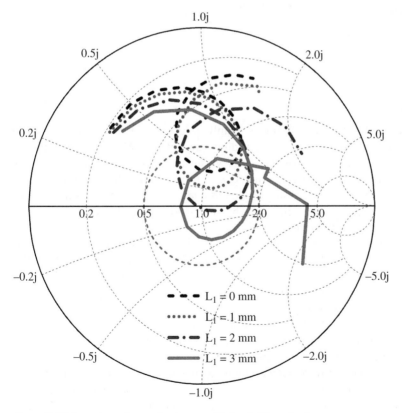

Figure 3.47 Input impedance of antennas with L_1 of different values.

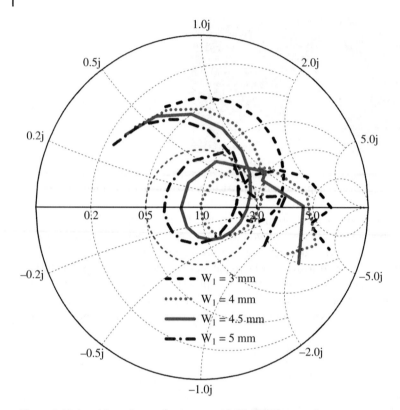

Figure 3.48 Input impedance of antennas with W_1 of different values.

the overlapped patch (case $L_1 = 0$ mm) is mainly inductively and the majority of the impedance loci falls outside the voltage standing wave ratio VSWR $= 2$ circle, indicating that the bandwidth at this state is very narrow. The inductive input impedance stems from the inner pin of the coaxial line, which goes through the substrate and introduces some inductance. The ring-shaped phase delayed line also contributes to some inductance. By increasing L_1, the impedance locus begins to move along the admittance circle due to the increased capacitance of C_1 and C_2.

Figure 3.48 shows the input impedance loci with different patch widths, W_1, when patch length L_1 is fixed to 3 mm. Similar to the results observed in Figure 3.47, the impedance loci moves along the admittance circle when W_1 becomes larger. This impedance matching procedure is similar to the method that uses lumped LC elements to tune the impedance bandwidth.

The distance between the diodes and the phase centre of the dipoles is represented by the geometry parameters R_4. The variation of R_4 affects the mutual coupling between the crossed dipoles and the parasitic rings, which influences the far-field radiation. To give a better understanding of how R_4 affects the performance of the antenna, Figure 3.49 shows the simulated directivity and the FBR of the present antenna under different values of R_4. The directivity increases when R_4 increases while the FBR decreases. This is a design trade-off and the value of R_4 should be chosen based on the design requirements.

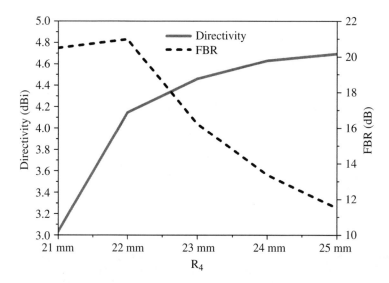

Figure 3.49 The simulated directivity and FBR under different values of R_4 at 2.3 GHz.

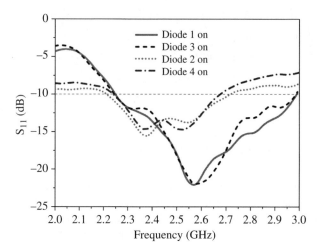

Figure 3.50 Measured reflection coefficient under four states of PIN diodes.

Figure 3.50 shows the reflection coefficient of the antenna. The measured results indicate that the designed antenna operates from 2.24 to 2.68 GHz when diode 2 or 4 is turned on and from 2.25 to 2.98 GHz when diode 1 or 3 is turned on. Notice that there are obvious differences between the results obtained when case diode 1 or 3 is on and case diode 2 or 4 is on. These differences are caused by the asymmetric structure of the antenna. Although the antenna is composed of four symmetric arms (driven element) and a rotational symmetric parasitic ring, the dual-layer structure of the driven element results in a different coupling magnitude.

Figure 3.51 shows the simulated and measured radiation patterns in the azimuth plane at 2.3 GHz. As can be seen in Figure 3.51, four directional beams can be obtained in the

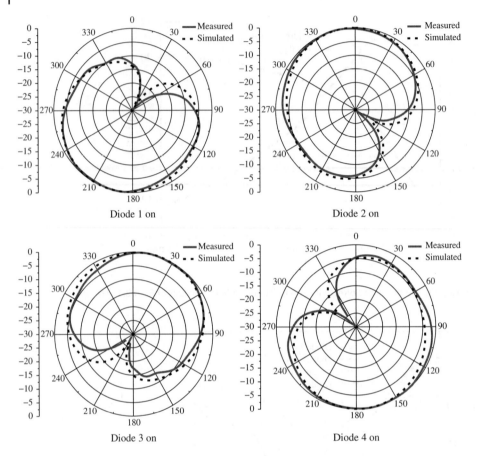

Figure 3.51 Simulated and measured radiation patterns in the azimuth plane at 2.3 GHz.

azimuth plane. The measured radiation patterns indicate that the antenna has a 3 dB beamwidth of about 120°, which makes this antenna able to provide the converge of the whole azimuth. In addition, the gain of the antenna is about 3.5 dBi under all states while the realised gain is around 3.85 dBi. Moreover, the measured FBR is larger than 20 dB.

3.7 Summary of the Chapter

In this chapter fundamental theories and design examples with case studies of EPSAR antennas are presented. It is shown that ESPAR provides a very low-cost solution for smart antennas, as it only has one active antenna element. There are no expensive phase shifters used in the ESPAR antenna and no complicated feed network is required. To realise good beamforming, it is important to optimise the reactance loads of each parasitic element. The link quality test proves that the EPSAR antenna improves the channel throughput and provides a reliable link for a wireless communication system even in a large noise environment.

References

1 D.V. Thiel and S. Smith. *Switched Parasitic Antennas for Cellular Communications*. Antennas and Propagation Library. Artech House, 2002.

2 R. Harrington. Reactively controlled directive arrays. *IEEE Transactions on Antennas and Propagation*, 26(3):390–395, May 1978.

3 C. Sun, A. Hirata, T. Ohira, and N.C. Karmakar. Fast beamforming of electronically steerable parasitic array radiator antennas: theory and experiment. *IEEE Transactions on Antennas and Propagation*, 52(7):1819–1832, July 2004.

4 T. Ohira and J. Cheng. *Analog Smart Antennas*, pages 184–204. Springer, Berlin, Heidelberg, 2004.

5 Skyworks. Varactor SPICE models for RF VCO applications. Technical report, Skyworks Solutions, Inc, 2015.

6 Skyworks. SMV1702-011LF hyperabrupt junction tuning varactor. Technical report, Skyworks Solutions, Inc, 2011.

7 R. Schlub, J. Lu, and T. Ohira. Seven-element ground skirt monopole ESPAR antenna design from a genetic algorithm and the finite element method. *IEEE Transactions on Antennas and Propagation*, 51(11):3033–3039, Nov 2003.

8 J. Cheng, Y. Kamiya, and T. Ohira. Adaptive beamforming of ESPAR antenna using sequential perturbation. In *2001 IEEE MTT-S International Microwave Sympsoium Digest (Cat. No.01CH37157)*, volume 1, pages 133–136, May 2001.

9 K. Iigusa, J. Cheng, and T. Ohira. A stepwise recursive search algorithm for adaptive control of the electronically steerable parasitic array radiator antenna. In *2002 32nd European Microwave Conference*, pages 1–4, Sept 2002.

10 H. Kawakami and T. Ohira. Electrically steerable passive array radiator (ESPAR) antennas. *IEEE Antennas and Propagation Magazine*, 47(2):43–50, April 2005.

11 J. Cheng, M. Hashiguchi, K. Iigusa, and T. Ohira. Electronically steerable parasitic array radiator antenna for omni- and sector pattern forming applications to wireless ad hoc networks. *IEE Proceedings – Microwaves, Antennas and Propagation*, 150(4):203–208, Aug 2003.

12 J. Lu, D. Ireland, and R. Schlub. Dielectric embedded ESPAR (DE-ESPAR) antenna array for wireless communications. *IEEE Transactions on Antennas and Propagation*, 53(8):2437–2443, Aug 2005.

13 E.E. Altshuler and T.H. O'Donnell. An electrically small multi-frequency genetic antenna immersed in a dielectric powder. *IEEE Antennas and Propagation Magazine*, 53(5):33–40, Oct 2011.

14 I. Ida, J. Sato, H. Yoshimura, and K. Ito. Improvement in efficiency-bandwidth product in small dielectric loaded antennas. *Electronics Letters*, 36(10):861–862, May 2000.

15 H. Liu, S. Gao, and T.H. Loh. Small director array for low-profile smart antennas achieving higher gain. *IEEE Transactions on Antennas and Propagation*, 61(1):162–168, Jan 2013.

16 H.T. Liu, S. Gao, and T.H. Loh. Electrically small and low cost smart antenna for wireless communication. *IEEE Transactions on Antennas and Propagation*, 60(3):1540–1549, March 2012.

17 H. Liu. *Design and Measurement Methodologies of Low-Cost Smart Antennas*. PhD thesis, University of Surrey, 2012.

18 K. Kabalan, A. El-Hajj, A. Chehab, and E. Yaacoub. Intercell interference reduction by the use of Chebyshev circular antenna arrays with beam steering. In *2007 National Radio Science Conference*, pages 1–7,March 2007.

19 H. Liu, S. Gao, and T.H. Loh. Compact dual-band antenna with electronic beam-steering and beamforming capability. *IEEE Antennas and Wireless Propagation Letters*, 10:1349–1352, 2011.

20 Y.G. Kim and N.C. Beaulieu. Binary grassmannian weightbooks for MIMO beamforming systems. *IEEE Transactions on Communications*, 59(2):388–394, February 2011.

21 S. Granieri, M. Jaeger, and A. Siahmakoun. Multiple-beam fiber-optic beamformer with binary array of delay lines. *Journal of Lightwave Technology*, 21(12):3262–3272, Dec 2003.

22 H. Liu, S. Gao, T.H. Loh, and F. Qin. Low-cost intelligent antenna with low profile and broad bandwidth. *IET Microwaves, Antennas Propagation*, 7(5):356–364, April 2013.

23 Atmel. At86rf231. Technical report, Microchip Technology Inc., September 2009.

24 Approved IEEE Draft Amendment to IEEE Standard for Information Technology-Telecommunications and Information Exchange Between Systems – Part 15.4: Wireless Medium Access Control (MAC) and Physical Layer (PHY) Specifications for Low-Rate Wireless Personal Area Networks (LR-WPANS): Amendment to Add Alternate Phy (amendment of IEEE Std 802.15.4). *IEEE Approved Std P802.15.4a/D7, Jan 2007*, 2007.

25 S. Zhang, G.H. Huff, J. Feng, and J.T. Bernhard. A pattern reconfigurable microstrip parasitic array. *IEEE Transactions on Antennas and Propagation*, 52(10):2773–2776, Oct 2004.

26 X.S. Yang, B.Z. Wang, S.H. Yeung, Q. Xue, and K.F. Man. Circularly polarized reconfigurable crossed-Yagi patch antenna. *IEEE Antennas and Propagation Magazine*, 53(5):65–80, Oct 2011.

27 S. Jeong and W.J. Chappell. A city-wide smart wireless sewer sensor network using parasitic slot array antennas. *IEEE Antennas and Wireless Propagation Letters*, 9:760–763, 2010.

28 A. Miura, W. Luo, M. Taromaru, M. Ueba, and T. Ohira. Experimental study of reactively loaded parasitic microstrip array antenna for circular polarization. In *The Second European Conference on Antennas and Propagation*, EuCAP 2007, pages 1–5, Nov 2007.

29 M.R. Nikkhah, P. Loghmannia, J. Rashed-Mohassel, and A.A. Kishk. Theory of ESPAR design with their implementation in large arrays. *IEEE Transactions on Antennas and Propagation*, 62(6):3359–3364, June 2014.

30 R.J. Mailloux. *Phased Array Antenna Handbook*. Artech House, Inc., 2005.

31 J.J. Luther, S. Ebadi, and X. Gong. A microstrip patch electronically steerable parasitic array radiator (ESPAR) antenna with reactance-tuned coupling and maintained resonance. *IEEE Transactions on Antennas and Propagation*, 60(4):1803–1813, April 2012.

32 C. Gu, S. Gao, H. Liu, Q. Luo, T.H. Loh, M. Sobhy, J. Li, G. Wei, J. Xu, F.Qin, B. Sanz-Izquierdo, and R.A. Abd-Alhameed. Compact smart antenna with electronic beam-switching and reconfigurable polarizations. *IEEE Transactions on Antennas and Propagation*, 63(12):5325–5333, Dec 2015.

33 R.H. Byrd, S.L. Hansen, J. Nocedal, and Y. Singer. A stochastic quasi-Newton method for large-scale optimization. *SIAM Journal on Optimization*, 26(2):1008–1031, 2016.

34 L. Zhang, S. Gao, Q. Luo, P.R. Young, and Q. Li. Planar ultrathin small beam-switching antenna. *IEEE Transactions on Antennas and Propagation*, 64(12):5054–5063, Dec 2016.

4

Beam-Reconfigurable Antennas Using Active Frequency Selective Surfaces

4.1 Introduction

Over the past 20 years, advances in metamaterials have opened up a new era for electromagnetic science and engineering. As a class of artificial structures, meta-materials can be engineered to modify electromagnetic properties that cannot be found in nature. A metamaterial consisting of subwavelength unit cells is essentially a resonant structure, and it is equivalent to a homogeneous material with effective relative permittivity and permeability. Thus, by tailoring these parameters, one can achieve exotic electromagnetic properties such as inversion of the Snell law, inversion of the Doppler effect, backward Cherenkov radiation, etc. Among the metamaterials developed so far, 2D metamaterial structures, usually termed meta-surfaces, have found many applications in microwave and antenna design because their planar periodic structures can be easily formed by arrays of metallic elements or apertures printed on dielectric substrates. Two typical examples are frequency selective surfaces (FSSs) and electromagnetic bandgap (EBG) structures [1].

2D EBG structure is a variant of the photonic bandgap (PBG) material. Photonic crystals in PBG materials are arranged periodically in high permittivity materials to prevent the propagation of electromagnetic waves in the optical region. To obtain EBG properties in microwave frequencies, 2D planar periodic structures are employed to produce electromagnetic bandstop and bandpass characteristics when excited by electromagnetic waves at an angle arbitrary to the plane of the array. Due to the low profile of the 2D EBG surfaces, they are attractive for integration with antennas compared to 3D dielectric woodpile and rod structures [2–5]. Therefore, there has been tremendous research work in the RF/antenna field, including that on high impedance surfaces (HIS), pioneering work introduced by Daniel F. Sievenpiper in [6]. The two unique properties of HIS are the in-phase reflection of incident plane waves and the surface wave suppression for any incident angles and polarisation states. A typical HIS is realised using mushroom-type metallic elements in a 2D periodic arrangement printed on a grounded dielectric substrate, as shown in Figure 4.1.

Compared with EBG, FSSs are easier to fabricate because they only consist of a 2D periodic array of patch or aperture elements without using any vias. In the past few decades, FSSs have found considerable application in antennas, filters, polarisers, radomes, and electromagnetic shielding due to their frequency filtering property and impact on the amplitude and phase of the incident waves. One of the main features of FSSs is that they can operate as bandstop or bandpass filters once exposed to

Low-cost Smart Antennas, First Edition. Qi Luo, Steven (Shichang) Gao, Wei Liu, and Chao Gu.

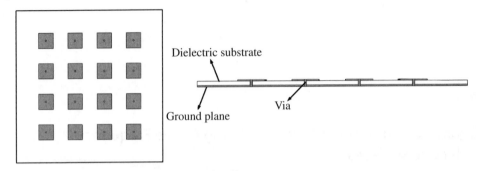

Figure 4.1 Top and side views of the mushroom-type HIS.

electromagnetic radiation [7]. Previous research has focused on the improvement of their functionality by decreasing their required size and sensitivity to the angle of incidence, developing multi-frequency selectivity, etc. FSSs can be composed of passive or active elements. Recently, the capabilities of FSSs have been extended by embedding active devices into the unit cell of the periodic structures. It is also possible to incorporate devices that provide gain or nonlinearity into FSSs, leading to the development of arrays with features such as oscillation, amplification, and mixing [8]. As the subject of interest in this chapter, active FSSs are capable of tuning the frequency response of the incident waves. By exploiting the reconfigurability of the active FSS, a class of FSS-based low-cost smart antennas can be developed with more design freedoms. This chapter discusses the techniques for designing electronically beam-switching or beam-steerable antennas using FSSs. These antennas do not require any RF phase shifters as in traditional phased arrays. Thus a significant reduction in the weight, power consumption, complexity, and cost of smart antennas can be achieved. This chapter starts with an introduction to FSSs including their basic principles and analysis techniques. This is followed by the discussions of several types of beam-switching or beam-steerable antennas using FSSs, including a single-band beam-switching antenna, a dual-band beam-switching antenna, a 3D beam-scanning antenna, and a frequency-agile beam-switching antenna. Details of antenna designs and the biasing techniques are explained, and many results are presented. A case study is also given in the last section of the chapter to provide the step-by-step design procedure of a beam-switching antenna using FSSs.

4.2 FSS Fundamentals and Active FSSs

This section discusses the basics of FSSs, including the element shape and dielectric loading effects. The analysis of an FSS is theoretically explained with an equivalent circuit. The concept of the active FSS and related biasing techniques are also presented.

4.2.1 FSS Elements

The geometry of the element is fundamental in defining the characteristics of an FSS. In general, aperture-type elements are used to produce a transmission band response,

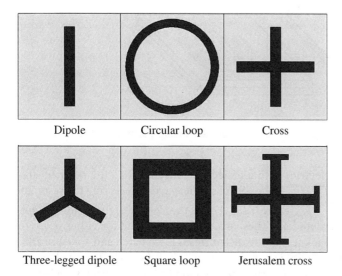

Dipole   Circular loop   Cross

Three-legged dipole   Square loop   Jerusalem cross

Figure 4.2 Different geometries of the FSS unit cells.

while bandstop performance can be obtained by applying resonant patch elements in the FSS. When designing either a bandpass or a bandstop FSS, the suitable element must be selected to meet the desired bandwidth and cross-polarisation level. The most common FSS element shapes are straight dipole, circular loop, cross dipole, three-legged dipole, square loop, and Jerusalem Cross, as shown in Figure 4.2.

Different types of FSSs show different responses. For example, some FSS elements show low cross-polarisation and less sensitivity to the polarisation states but with narrow bandwidth, while some elements are inherently broadband but with a higher cross-polarisation level. The resonant frequency of the freestanding dipole element array has the worst stability with incident angle variations. If the incident direction is oblique to the broadside of the dipole, the dipole cannot resonate effectively because the projected length of the dipole is less than a half-wavelength. Therefore, the operating frequency shifts depending on the angle of incoming waves. Loop elements provide broad bandwidth and stable response with the angle of incident and polarisations. For square-loop and circular-loop elements, resonance occurs when the length of the whole loop is approximately one wavelength. If the element size is different from the resonant dimensions, the FSS screen is transparent to the RF signal and the microwave wave can pass through it with some insertion loss. The insertion loss is due to the dielectric, copper conduction, and scattering loss. In short, having an element with compact size and stable response to the different incident angles is the most important criteria in element selection. Small element size also leaves the designer more design freedom to choose an appropriate lattice pattern for the intended application.

4.2.2 Dielectric Loading Effects

In general, an FSS is backed by dielectrics and there are two basic dielectric loading techniques: (i) the elements are bonded in one side of the dielectric and (ii) the elements are sandwiched between two dielectrics, as shown in Figure 4.3. If the FSS is

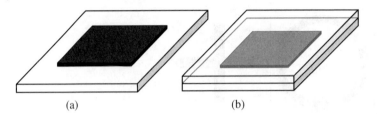

(a)   (b)

Figure 4.3 Two different types of FSS configuration: (a) printed on single dielectric substrate and (b) sandwiched between two dielectrics.

sandwiched between two dielectrics with a dielectric constant of ε_r, the resonant frequency decreases from f_0 to $f_0/\sqrt{\varepsilon_r}$. If the FSS has a dielectric on one side the resonant frequency shifts to $f_0/\sqrt{(\varepsilon_r + 1)/2}$, where $(\varepsilon_r + 1)/2$ is the average of the dielectric constant of the dielectric material and air. These expressions are valid for either patch or aperture (slot) arrays. If the thickness of the dielectrics is high, patch and aperture FSSs behave differently. Angular stability is lost for slot arrays if the dielectric thickness is increased but is regained if the dielectric thickness is a multiple of a quarter-wavelength.

4.2.3 FSS Analysis Techniques

Since the 1960s, several methods have been proposed to analyse FSSs. One method is the equivalent circuit method (ECM). The ECM is a simple approximation method and is relatively fast [9]. In this technique, the FSS periodic array is modelled as a circuit with inductive or capacitive components. Based on the solution of the resistor-inductor-capacitor (RLC) circuit, the reflection and transmission coefficients of the FSS can be calculated. Given that this method uses a quasi-static approximation to calculate the circuit components, it is only accurate at the resonant frequency of the FSS. One advantage of this method is that very little computing power is required. The ECM assumes that the arrays are periodic with infinite size, and this method is limited to the normal incident angle. As shown in Figure 4.4, the patch element is modelled as a parallel circuit of a capacitor and an inductor, and the array of aperture elements is represented by a series LC circuit, where the conducting element provides the inductance and the inter-element spacing provides the capacitance. From the equivalent model, the resonant frequency can be calculated

$$f_0 = 1/(2\pi\sqrt{LC}) \tag{4.1}$$

The fractional bandwidth (BW) is defined as the difference between the lower and upper frequency divided by the centre frequency, and BW's relationship with the inductance and capacitance of the circuit is given by the following equation

$$BW \propto \sqrt{L/C} \tag{4.2}$$

Thus, increasing the length of the conducting element, which is equivalent to increase the inductance, can decrease the resonant frequency and improve the bandwidth. The resonant frequency can also be decreased by increasing the capacitance but this results in a narrower bandwidth.

Another analysis method is modal analysis. In this method, the distribution of current induced in conducting elements (or fields in slots) is represented as a series of

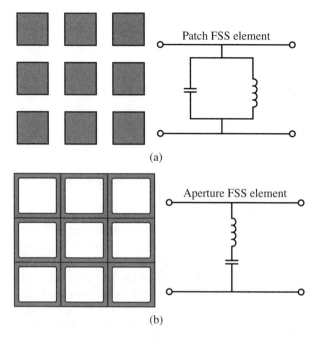

Figure 4.4 FSS using (a) patch and (b) aperture elements and their equivalent circuit models.

suitable basis functions (usually sinusoidal functions or waveguide modes). The method involves the derivation of the integral equation by matching the Floquet modes [10] in space and the aperture or current modes on the periodic surface. The integral equation is formulated using the spatial or spectral-domain approach. Using the spectral-domain approach helps to reduce the complexity of the integral equations to a simple algebraic multiplication of simple functions. The method of moments and the conjugate gradient techniques are the most used methods to solve the integral equation.

The third FSS analysis technique is the iterative process [11]. The first estimation of the element current induced by the incident field E_{inc} is used to calculate the local electric field distribution E, which is then set to zero at the surface of the conductors. This modified field is used to recalculate the element current distribution. The root mean square (RMS) value of E/E_{inc} over the conductors can be used as an indicator of the iteration progress. The first two of these techniques assumes the FSS array is periodic and infinite in size. The third method is not constrained by these two conditions and variants [12] are used to study the edge effects and the element currents across non-uniformly illuminated FSSs of finite size [13]. An in-depth review of analysis techniques is given by Mitra et al. [14]. Other analytical methods include the finite difference time domain (FDTD) method, hybrid methods, and finite element methods. Optimisation methods such as genetic algorithms can be used to assist the design of an FSS to meet specific requirements [15–18].

4.2.4 Active Metal Strip FSS

The starting point to develop an active FSS is the equivalent circuit representation of an infinite array of parallel conducting strips. The transmission response of the conducting

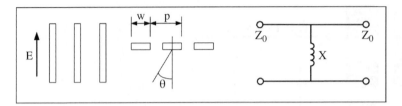

Figure 4.5 Equivalent circuit model of the inductive strip FSS.

strips array is frequency dependent. It also depends on the orientation of the incident electric field relative to the strips.

By using the analogy with a transmission line of characteristic impedance equal to that of free space, the strip FSS can be modeled as a lumped reactance circuit. The analysis of the strip array formulation for transverse electric (TE) incidence is shown in Figure 4.5. The metal strips have width w and period p. The plane wave is incident onto the strips with angle θ. When the tangential component of the incident electric field is parallel to the conductors, this reactance is inductive, allowing very little transmission in low frequencies where the inductance virtually shorts the line. The value of the inductance depends not only on the p and w, but also on the incidence angle θ and the mode of the incidence wave (TE or transverse magnetic, TM). The inductive reactance of the equivalent circuit is calculated by [19]

$$\frac{X(w)}{Z_0} = F(p, w, \lambda) = \frac{p \cos \theta}{\lambda} \left\{ \ln \frac{1}{\sin\left(\dfrac{\pi w}{2p}\right)} + G(p, w, \lambda, \theta) \right\} \tag{4.3}$$

where Z_0 is the characteristic impedance of free space. G is the first-order correction terms for both the TE and TM waves and can be expressed by

$$G(P, W, \lambda, \theta) = \frac{1}{2} \frac{(1 - \beta^2)^2 \left[\left(1 - \dfrac{\beta^2}{4}\right)(A_+ + A_-) + 4\beta^2 A_+ A_- \right]}{\left(1 - \dfrac{\beta^2}{4}\right) + \beta^2 \left(1 + \dfrac{\beta^2}{2} - \dfrac{\beta^2}{8}\right)(A_+ + A_-) + 2\beta^6 A_+ A_-} \tag{4.4}$$

$$A_\pm = \frac{1}{\sqrt{\left(\dfrac{p \sin \theta}{\lambda}\right)^2 - \left(\dfrac{p}{\lambda}\right)^2}} - 1 \tag{4.5}$$

$$\beta = \sin \frac{\pi w}{2p} \tag{4.6}$$

Similarly, the equivalent circuit representation for TM incidence is shown in Figure 4.6. The incident magnetic field vector is parallel to the metal strips and is

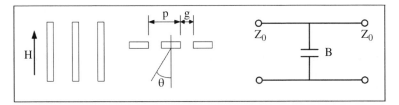

Figure 4.6 Equivalent circuit model of the capacitive strip FSS.

incident at an angle of θ. The capacitive susceptance is calculated by

$$\frac{B(g)}{Z_0} = 4F(p, g, \lambda) = \frac{4p \cos \theta}{\lambda} \left\{ \ln \frac{1}{\sin \left(\frac{\pi g}{2p} \right)} + G(p, w, \lambda, \theta) \right\} \qquad (4.7)$$

Note that Equations 4.3–4.7 are valid when $w \ll p$, $g \ll p$ and $p \ll \lambda$. The wavelengths and angles of incidence θ should be in the range $(1 + \sin \theta)/\lambda < 1$. They are also only valid for plane wave incident in either the E or H plane and hence they cannot be used to model the cross-polarisation effects of the FSS.

Based on the above discussion, tunable or reconfigurable planar FSS is developed by changing the equivalent impedance of the screen seen by the incident plane waves from an inductive state to a capacitive state. Figure 4.7 shows a typical strip-type active FSS unit cell incorporating a PIN diode. In this configuration, $w \ll P_x$, $P_x \ll \lambda$, and $g_0 \ll P_x$ are considered to extract the related parameters for the transmission lines. Moreover, since the strips are very thin compared to the wavelength, the thickness effect is not considered. For the ideal case, when the diode is activated, an equivalent microstrip

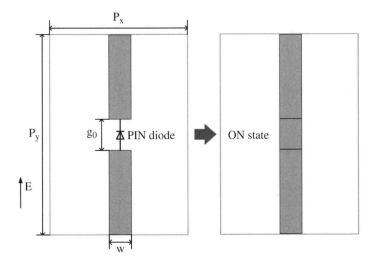

Figure 4.7 Active strip FSS when the PIN diode is in the ON state.

connects the two short strips. A high pass response is obtained for TE incident waves (polarised parallel to the strips), which is equivalent to the passive strip FSS shown in Figure 4.5.

According to the equivalent reactance value of an inductive surface calculated by Equation 4.3, the expression can be simplified for the normal incidence EM wave to the following equation

$$X_{TE} = F(P_x, w, \lambda) = \frac{P_x}{\lambda} \left[\text{Incosec} \left(\frac{\pi w}{2P_x} \right) + G(P_x, w, \lambda) \right] \tag{4.8}$$

When $P_x \ll \lambda$, the small correction term G can be neglected and this equation can be rewritten as

$$X_{TE} = \frac{P_x}{\lambda} \ln \left(\frac{2P_x}{\pi w} \right), w \ll P_x \ll \lambda \tag{4.9}$$

The transmission response of the inductive screen can be defined in terms of grid lattice dimensions. This equation demonstrates that the total inductance of the grid is directly proportional to the periodicity P_x. In addition, by increasing the width of strips, the logarithmic term in this equation reduces, which leads to less inductance for a given set of periodicities. Therefore, to increase the total inductance of the screen, P_x needs to be increased and w must be decreased.

Figure 4.8 illustrates the case when the diode is switched off. The diode is replaced by a capacitor and the whole structure shows a bandstop response where its frequency depends on the total length of the dipole-like FSS. The effective length of the resonant short strip dipole determines the inductance value and the gap is set to be small to accommodate the PIN diode package. Thus, the reactance of the screen is approximately the same as that of the inductive grid case and is given by

$$X_{TE} = \frac{P_y - g_0 P_x}{P_y} \frac{1}{\lambda} \ln \left(\frac{2P_x}{\pi w} \right) \cong \frac{P_x}{\lambda} \ln \left(\frac{2P_x}{\pi w} \right), w \ll P_x \ll \lambda \tag{4.10}$$

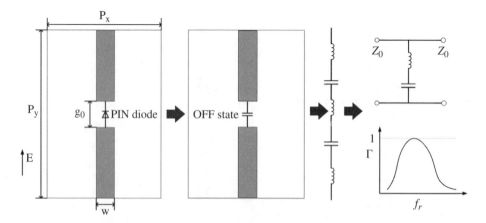

Figure 4.8 Active strip FSS when the PIN diode is in the OFF state.

The related susceptance is greatly affected by the effective strip width and can be expressed as

$$B = 4F(P_y, g_0, \lambda)\frac{w}{P_x} = 4\left(\frac{P_y}{P_x}\right)\left(\frac{w}{P_x}\right)\left(\frac{P_y}{\lambda}\right)\ln\left(\frac{2P_y}{\pi g_0}\right) \tag{4.11}$$

It can be seen from the above equation that $\frac{P_y}{P_x}$ is a critical parameter, which has a strong impact on the center frequency of the FSS bandstop.

4.2.5 Active Slot FSS

As the Babinet counterpart of the strip FSS, slot FSS can also be reconfigured using PIN diodes or varactors. The development of slot active FSS structures begins with passive slot FSS. Theoretically, the slot FSS consists of an infinitely thin conducting screen with arrays of dipole apertures and offers a bandpass frequency filtering response. The main difference between strip and slot FSSs is that electric currents are induced on the metal dipole for the strip while magnetic currents are excited for the slot.

As can be seen from Figure 4.9, an incident plane wave impinges on the filter, which causes the electrons in the metal to oscillate and a small portion of the energy radiates. When the frequency of the incident wave is in accordance with the resonant frequency of the slot, most of the energy is re-radiated because the electrical path of the electrons is about a half-wavelength and the slot radiates like a dipole. At higher frequencies, the electrons flow in a smaller space and the current paths are cut into several small ones, which leads to inefficient radiation. Base on the aforementioned equivalent circuit analysis, the slot FSS can be modeled as a capacitive parallel circuit. Figure 4.10 shows the frequency response with the incident wave perpendicular to the slot. At its resonant frequency, the FSS can be regarded as a transparent EM window.

To reconfigure a slot-type FSS, PIN diodes or varactors are incorporated in the centre of the slot, as can be seen from Figure 4.11. Here an ideal PIN diode is used to investigate the equivalent circuit model of the FSS structure. The surface impedance of the single

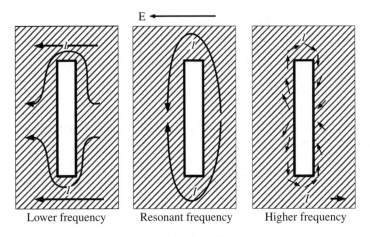

E ◄────────

Lower frequency   Resonant frequency   Higher frequency

Figure 4.9 Current distribution of the slot FSS.

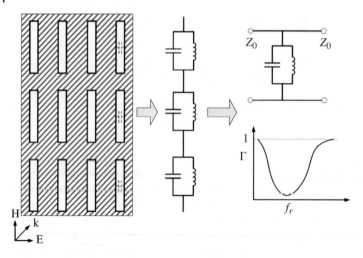

Figure 4.10 Equivalent circuit model of the slot FSS.

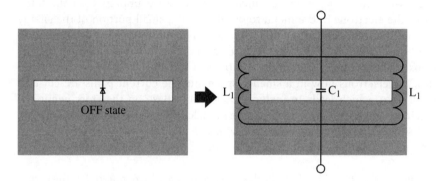

Figure 4.11 Slot-type active FSS when the PIN diode is in the OFF state.

slot model at diode OFF state can be expressed as

$$Z_{FSS} = \left(j\omega C_1 + \frac{2}{j\omega L_1} \right)^{-1} = \frac{j\omega C_1}{2 - L_1 C_1 \omega^2} \tag{4.12}$$

The resonant frequency of the parallel resonant circuit model is defined as

$$\omega_0 = \sqrt{\frac{2}{L_1 C_1}} \tag{4.13}$$

When the PIN diode is switched ON, as shown in Figure 4.12, the surface impedance of the approximate model is given as

$$Z_{FSS} = f(L_1, L_2, C_1, C_2) = \frac{-L_1 L_2 \omega^2}{2j\omega L_2 + j\omega L_1 - 2j\omega^3 C_2 L_1 L_2} \tag{4.14}$$

The transmission coefficient of the slot FSS, T, is calculated by

$$T = \frac{2Z_{FSS}}{2Z_{FSS} + \eta} \tag{4.15}$$

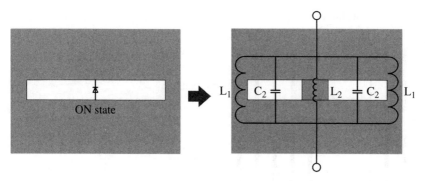

Figure 4.12 Slot-type active FSS when the PIN diode is in the ON state.

where η is the free-space intrinsic impedance. In the case where the diode is in the OFF state, the FSS has a bandpass response. Assume the 3 dB cut-off angular frequency is w_c, then

$$|T|_{w=w_c} = \frac{1}{\sqrt{2}} \tag{4.16}$$

Substituting (4.16) into (4.15), we can obtain L_1, C_1 in (4.12). Similarly, L_2, C_2 in (4.14) can also be calculated based on the two poles of the surface impedance of the FSS with the diodes switched ON.

4.2.6 Active FSS Biasing Techniques

To apply an external voltage to the diodes, DC bias networks need to be carefully designed as they can degrade the performance of the FSS when integrating with the antennas [20–22].

Figure 4.13 shows the two configurations of DC bias networks reported in [20], where diodes are connected along each row or each column. As the sizes of the bias lines are comparable to the FSS element size, introducing the DC bias lines causes a considerable decrease in the bandstop pole frequency. This can be explained by changing the total L and C values of the resonant circuit model. As a result, the loading effect of these lines undesirably affects the EM response of the unit cell, which causes some shifts in the frequency of the bandstop.

To avoid the unwanted effects from the bias circuit, it is better to print the FSS and the bias lines in different layers [23–26]. Figure 4.14 shows the configuration of the active FSS unit cell that consists of a double-sided structure, where the biasing circuit is isolated from the passive FSS elements. The slots and bias lines are printed on a flexible polyester substrate of 0.05 mm thick with metal cladding on its two sides. The FSS with the slots is on the top layer while PIN diodes are placed on the biasing circuit at the bottom layer, thereby hiding the lines from the incident EM waves. The active slot FSS is suitable for antenna applications such as beam steering antennas and reconfigurable EBG antennas because of the minimised signal perturbation. By using a ring or other FSS elements, reconfigurable multi-band FSSs can also be achieved, which provides more functionality when applied to antenna applications.

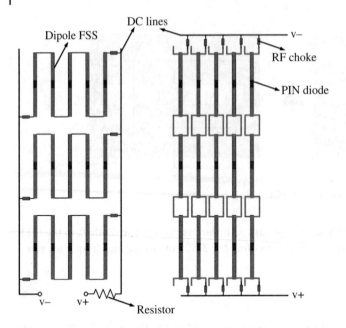

Figure 4.13 Two biasing configurations of the strip-type active FSS described in [20].

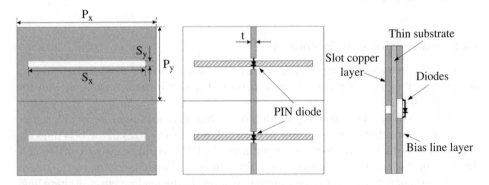

Figure 4.14 Bias technique of the slot-type active FSS reported in [26].

4.3 Monopole-fed Beam-switching Antenna using AFSS

This section introduces the design of a beam-switching antenna using an active FSS (AFSS). The operation principle is to employ conventional radiating sources, e.g. dipoles and monopoles, as the feed to illuminate the AFSS. Then through interacting with the FSS that is loaded with relatively low-cost off-the-shelf components, e.g. PIN diodes and varactors, the beam of the antenna can be reconfigured. In this way, a low-cost beam-switching antenna can be developed by combing a simple feed with multifunctional FSSs.

FSSs are essentially filters to electromagnetic waves. They can be designed to be either transmitting or opaque in the required frequency bands. By incorporating PIN

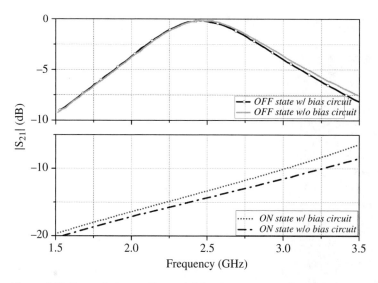

Figure 4.15 Transmission coefficients of the slot-type active FSS reported in [26] when the PIN diode is switched ON and OFF.

diodes, a bandpass slot FSS can be switched between transmitting and reflecting states when switching the diodes ON and OFF [26]. The unit cell of the active FSS shown in Figure 4.14 consists of passive FSS elements and a biasing circuit including a PIN diode, which were printed on different sides of a 0.05-mm thick flexible polyester substrate ($\varepsilon_r = 3$). The dimensions of the FSS unit cell are $P_x = 57$ mm, $P_y = 25$ mm, $S_x = 55$ mm, $S_y = 0.5$ mm, and $t = 2$ mm. The RLC model of the diode is used in the full-wave simulation. When the diode is forward biased (ON state), it has a resistance R_s of 2.1 Ω. When the PIN diode is at zero or reverse biased (OFF state), the equivalent model can be approximated as a capacitance C_T of 0.2 pF. It is noted that several sets of unit cell dimensions can result in the same resonant frequency, so the specific set of values should be optimised for the final AFSS design. Figure 4.15 shows the simulated transmission coefficients of the FSS unit cell when the PIN diode is switched ON and OFF. As shown, the bias circuit has little effect on the transmission response when the diodes are switched OFF but has more effect on the transmission coefficients when the diodes are forward biased.

The configuration of the resulting antenna incorporating a cylindrical AFSS (CAFSS) is shown in Figure 4.16. For better illustration, half of the CAFSS in the perspective view is set as transparent. As the radiating source, a monopole antenna is placed in the centre of a round metallic ground and is encircled by the CAFSS. A flat AFSS with 12 columns and eight rows is rolled to form the CAFSS, so there are in total 96 PIN diodes used. The CAFSS is divided into two sectors. When the diodes in one sector are switched OFF, those in the other sector are switched ON, and vice versa. At the designed frequency, the sector with diodes switched OFF is EM transparent, while the sector with diodes switched ON acts as a reflector. Thus, the omnidirectional beam of the monopole feed can be transformed to a directional beam. By switching the PIN diodes in an appropriate combination of columns, one can not only achieve the function of beam-sweeping in the

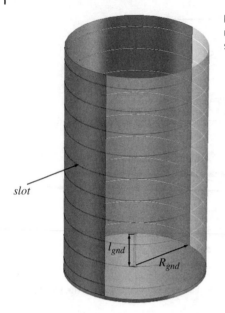

Figure 4.16 Perspective view of the CAFSS antenna reported in [26] (bias lines and PIN diodes are not shown for clarity).

entire azimuth plane but also adjust the beamwidth. The radius of the cylinder is

$$R_{CAFSS} = \frac{12 \times P_x}{2\pi} \tag{4.17}$$

The dimensions of the unit cell and the monopole feed structures are optimised to achieve the optimum antenna performance at different modes. All the design parameters are shown in Figure 4.16, where R_{gnd} is the radius of the ground as well as the CAFSS, while l_{ant} and r_{ant} are the length and radius of the monopole, respectively.

The radiation performance of the antenna depends on the overall dimension of the CAFSS and ground plate. Therefore, R_{gnd} is optimised to determine the horizontal tangential periodicity P_x. A metallic reflector with radius R_{gnd} is used in the EM simulation to examine the effects of the two key design parameters, P_x and N. Here N is the number of columns with PIN diodes in the OFF state. As shown in Figure 4.17, the directivity peaks at $N = 3$, regardless of the value of P_x. However, as can be seen from Figure 4.18, the antenna input impedance is mismatched at the desired band when P_x = 55 mm but the maximum directivity is 13 dBi. The larger P_x is, the better the antenna impedance matching, but the lower the directivity. Therefore, there is a trade-off between impedance matching and antenna directivity, which is determined by P_x and N. Taking into account all the antenna radiation performance, the optimal value of P_x for WiFi applications is chosen to be 57 mm.

In order to evaluate the performance of the antenna, the structure in Figure 4.16 is simulated using full-wave software CST Microwave Studio. A very thin polyester substrate is used with a thickness of 0.05 mm. This thickness is small compared to the wavelength at 2.45 GHz. Thus, the meshing process of such a thin cylindrical structure increases the simulation time significantly, making the antenna optimisation time-consuming and impractical. Indeed, the thin polyester substrate only has a minor effect on the resonant frequency of the FSS and contributes little to the antenna radiation

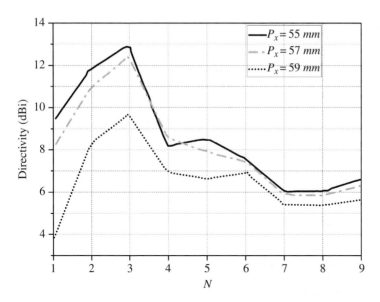

Figure 4.17 Effect of N and P_x on the directivity of the metallic reflector.

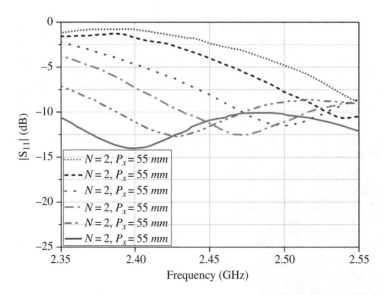

Figure 4.18 Effect of N and P_x on impedance matching of the metallic reflector.

characteristics, thus it is recommended not to include substrate thickness in the simulation.

Figure 4.19 shows the simulated H-plane radiation patterns for the two biasing states. It is clear that the antenna has a narrower beam pattern when $N = 3$, while a wide beam pattern can be obtained when $N = 7$. These results agree with the above discussion based on the equivalent metallic reflector. A prototype was fabricated and measured. A commercially available monopole Cushcraft SM2403M was used as the feed. A planar

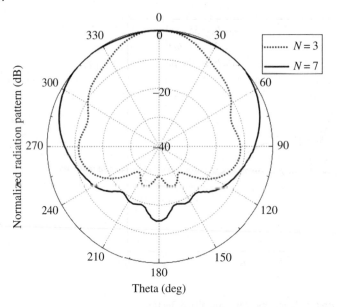

Figure 4.19 Directional and broad beam modes of the antenna reported in [26].

AFSS consisting of 12 × 8 unit cells was fabricated and tested. Twelve parallel biasing lines were etched on one side of the substrate, each of which incorporated eight PIN diodes (Infineon BAR 64-02). At the end of each bias line, a resistor of 1 kΩ was used to limit the current of the diodes. The assembled antenna prototype is shown in Figure 4.20.

For the OFF state, zero DC voltage was applied to the corresponding columns. A DC voltage of 6 V was applied to the column with diodes in the ON state. The simulated and

Figure 4.20 Assembled CAFSS antenna prototype.

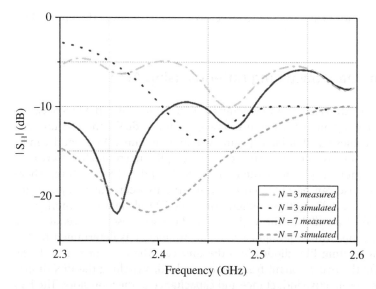

Figure 4.21 S_{11} of the two operating modes of the CAFSS antenna.

measured S_{11} results of the antenna are shown in Figure 4.21. The CAFSS antenna has reflection coefficients from -5 to -10 dB over the operating band of 2.4–2.5 GHz in the high-gain mode. For the low-gain mode, better impedance matching can be obtained. The degradation of the antenna impedance matching compared with the simulated one is mainly caused by the fabrication inaccuracies.

A measured gain of more than 6 dBi can be achieved for the narrow-beam mode and higher than 4 dBi gain for the wide-beam mode in the entire operating band. Two switched beams for both modes are demonstrated in Figure 4.22. Using the same principle and control of the states of the PIN diodes in corresponding columns, the

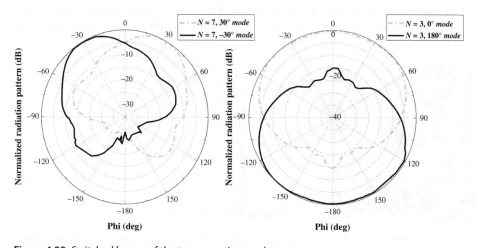

Figure 4.22 Switched beams of the two operating modes.

two beam-sweeping modes can generate consecutive beams to cover the entire 360° azimuth plane.

4.4 Dual-Band Beam-Switching Antenna using AFSS

The key to realising an FSS-based beam-switching antenna is to switch the FSS between transmitting and reflecting modes. Therefore, a reconfigurable dual-band FSS unit cell can be designed to achieve a dual-band beam-switching antenna. As a good candidate, active slot-element FSSs comprising two concentric split-ring resonators were proposed in [24]. The authors in [27] presented a new dual-band AFSS structure, where a dual-bandpass response can be obtained by tuning the two concentric square slot rings, as shown in Figure 4.23. In this configuration, the EM plane waves impinge on the FSS with the E-field parallel to the y direction. The lower resonant frequency is related to the outer loop, while the higher resonant frequency is determined by the inner loop. By incorporating PIN diodes into the unit cell, the two concentric loops are split, which shifts the two resonant frequencies. Indeed, switching the PIN diodes ON or OFF changes the effective inductance and capacitance of the loop slots. The FSS elements are etched on 0.05 mm thick flexible substrate with $\varepsilon_r = 2.8$. This thin substrate is used to form the planar FSS into a cylinder. As can be seen from the back view of the unit cell, four PIN diodes are placed in different positions for each slot. When these diodes are forward or reverse biased, in theory the FSS sheet can be reconfigured as transparent or opaque at its resonant frequency. The BAR 64-02V silicon PIN diode was used and it was modelled as a resistor with resistance $R_s = 2.1\ \Omega$ ($0.85\ \Omega \leq R_s \leq 2.8\ \Omega$) for the ON state. A parallel connection of a capacitance $C_s = 0.17$ pF and a resistance $R_{off} = 5$ kΩ was used to model the PIN diodes in the OFF state. To obtain the required dual-bandpass responses, the related design parameters of the element geometry should be optimised, and the final dimensions of the resulting AFSS unit cell are as follows: $p = 17$ mm, $b = 11.8$ mm, $g = 1$mm, $a = 8.2$ mm, $l = 8.2$ mm, $wf_1 = 1$ mm, $wf_2 = 1$ mm. Note that the periodicity p of the square unit cell is the same for both x and y directions.

The passive and active FSSs are simulated, and the computed magnitude of transmission coefficients are shown in Figure 4.24. Three types of configurations are investigated

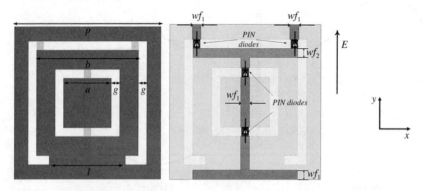

Figure 4.23 Configuration of the AFSS unit cell reported in [27]: front view with two concentric slots and back view with the bias circuit.

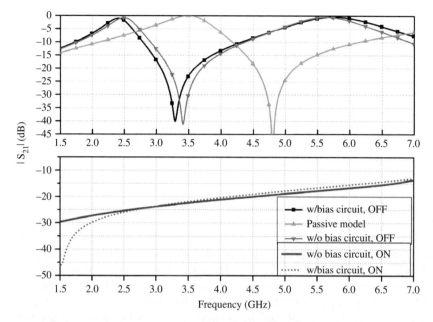

Figure 4.24 Simulated transmission coefficients of the passive and active FSS in [27].

with the same geometrical dimensions. It can be seen that by loading the four OFF state PIN diodes, the bandpass of the passive FSS is significantly shifted to lower frequencies due to the added capacitance between the gaps of the two slots. To study the effect of the bias circuit, the feed lines are removed whereas diodes are kept between the gaps of the passive element. It is observed that the bias circuit has a minor effect on the transmission responses. At normal incidence, the presented AFSS resonates at 2.5 GHz and 5.5 GHz when all diodes are OFF, where insertion loss higher than 15 dB between 2 and 6 GHz is achieved with the diodes in the ON state.

Besides the geometrical dimensions of the unit cell, the electrical characteristics of the PIN diodes are also very important. It is described in [27] that R_s has little impact on the magnitude of the transmission coefficient whereas a small change of diode capacitance leads to a significant shift of the bandpass poles. The very thin substrate (0.05 mm) is found to have almost no effect on the AFSS transmission responses. Figure 4.25 shows the 3D structure of the dual-band AFSS antenna. As can be seen, the antenna is composed of a reconfigurable cylindrical FSS enclosing a dual-band dipole. The FSS structure consists of unit cells, as shown in Figure 4.23. Ten columns of AFSS are employed and placed cylindrically with an angular periodicity of $\theta_{fss} = 36°$ and a radius of $R_{fss} = 27$ mm. Each column has 12 unit cells, and the total length of the antenna L_{fss} is 204 mm. For figure clarity, some columns are shown as transparent.

The dual-band dipole feed antenna is placed along the axis at the center of the cylindrical FSS. To transform the omnidirectional pattern into a directional one, ten columns (parallel to the cylinder axis) are divided into two sectors. In this way, four beams are generated to cover the horizontal (x–y) plane in 90° steps. In each step, the diodes of one sector are OFF while those in the other sector are ON. The semi-cylinder sector with ON-state diodes reflects the incident EM wave while the sector with OFF-state

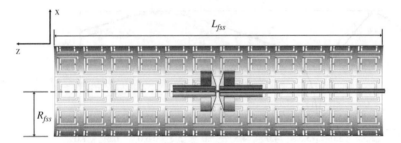

Figure 4.25 Cross-section view of the AFSS antenna reported in [27] (substrate and bias lines are not shown in the model).

diodes is transparent to the EM wave. By switching ON/OFF the diodes of each sector, the antenna radiation pattern can be swept in four steps with the same bias configurations because of the symmetrical structure. The key to realising a directional beam at the two operating frequencies is to optimise the overall size of the metallic semi-cylindrical reflector illuminated by the dual-band dipole. From the theory of the corner reflector, the distance of the dipole to the apex of the reflector R_{fss} substantially determines the antenna radiation pattern, including the side and back lobe level. Therefore, it is important to optimise R_{fss} in order to obtain the desired antenna performance at the two frequency bands.

For the prototyping, the planar AFSS is printed on a Kapton substrate and then conformed to a cylinder. Based on the designed unit cell period P, the number of the unit cells along the circle circumference can be calculated using $N = 2 \times \pi \times R_{fss}/P$ and the unit-cell number along the reflector length can be calculated by $M = L/P$. In this design, with $N = 10$ and $M = 12$, the AFSS antenna can have comparable performances as the dual-band semi-cylindrical reflector antenna. For all the beam-switching modes, five out of ten columns of PIN diodes need to be forward biased. According to the measurement setup, switching on each of the five parallel columns requires a 35 V DC power supply. The DC flowing in each column is then 20 mA. Thus, the total power consumption for each case is $P = 35 \times (0.02 \times 5) = 3.5$ watts.

The performance of the fabricated planar AFSS was characterised by measuring the transmission coefficients. The AFSS consists of 10×12 unit cells. There are ten parallel bias lines with RF chokes (PANASONIC ELJQE3N9ZFA) added to one end. At the other end, the bias lines are connected to a common line and this line is attached to the ground, which consists of the slot FSS copper layer. A 1 kΩ resistor is added to limit the current to the diodes. Figure 4.26 shows the measured transmission coefficients of the AFSS when all the diodes are switched ON and OFF. It can be seen that the frequency responses of the two states agree well with the simulation results except for some small discrepancies in the higher band, which is caused by the diode package, substrate losses and measurement inaccuracies. The measurement inaccuracies are related to the small FSS array size and small isolation screen aperture used for the tests as well as the non-plane wave characteristics of the transmitted signal. Figure 4.27 shows the fabricated cylindrical AFSS antenna prototype. It has dimensions of approximately $54 \times 54 \times 204$ mm.

During the measurements, five adjacent columns of diodes are OFF and the others are ON. The OFF state requires a DC voltage of 0 V, and a 35 V DC power supply is applied to the five parallel bias lines to activate the diodes. Figure 4.28 compares the

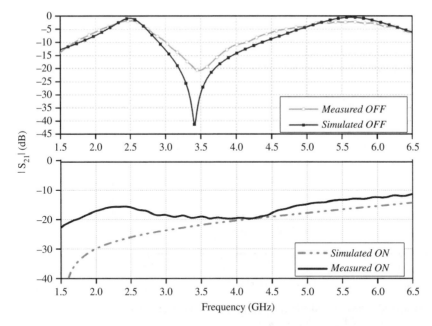

Figure 4.26 Measured and simulated transmission coefficients when diodes are OFF/ON.

Figure 4.27 Photograph of the fabricated antenna.

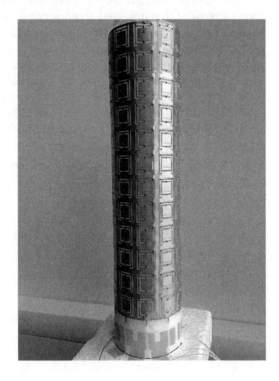

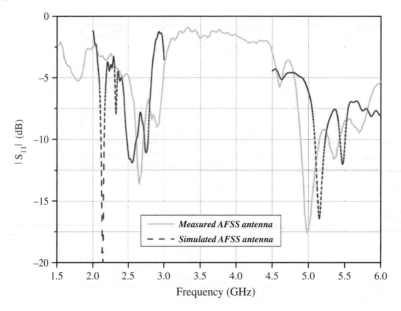

Figure 4.28 Measured and simulated reflection coefficients of the AFSS antenna.

simulated and measured reflection coefficients of the AFSS antenna. The measured radiation patterns of the presented antenna are shown in Figures 4.29 and 4.30. The 3 dB H-plane beamwidths of the two operation bands are 90° and 100°, respectively. The highest backlobe level at lower frequencies is about −18 dB while it is −12 dB at the higher frequencies. The normalised E-plane patterns show a minimum sidelobe level of −12 dB at 2.5 GHz. Moreover, the cross-polarisation levels within the 3 dB beamwidths of the two planes are below −12 dB. It can be seen that the main beams of the higher band have 20° beam tilt from the horizontal plane, which is caused by placing the coaxial feed cable asymmetrically in the cylindrical AFSS. The length of the cable is more than one

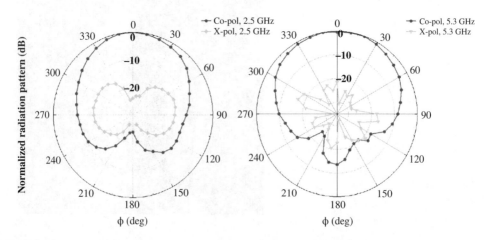

Figure 4.29 Measured H-plane radiation patterns of the antenna reported in [27].

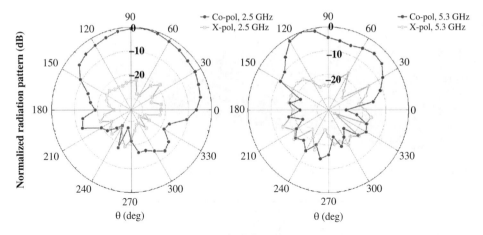

Figure 4.30 Measured E-plane radiation patterns of the antenna reported in [27].

wavelength at 5GHz, which deteriorate the E-plane radiation pattern drastically at the higher operation band. This distortion can be reduced by using the absorbing material. However, this would lead to a reduction in antenna radiation efficiency. The gain was measured in the plane of the maximum radiation. The measured gain results show that the AFSS antenna has gains of 6.5 dBi and 5 dBi at the two desired frequencies. Moreover, it is demonstrated in [27] that the radiation patterns can be switched in four steps to cover the whole azimuth plane at both frequencies.

4.5 3D Beam Coverage of Electronic Beam-Switching Antenna using FSS

An active FSS fed by beam-scanning collinear antenna arrays was reported in [28]. This antenna can achieve simultaneous beam-switching in the horizontal plane and beam-tilting in the elevation plane. The concept of steering the antenna main beam in the entire azimuth plane is presented in section 4.3, where a conventional monopole antenna is used as the feed. In order to realise beam-steering in the other plane, a customised antenna array feed should be used. The design concept is inspired by the base station antenna design, which consists of an incorporated feed dipole array over a planar metallic reflector [29, 30]. It is expected that by exciting the dipole array with specific amplitude and phase distribution, beam-scanning in the elevation plane can be achieved. As the starting point, a two-element dipole array is designed. As can be seen in Figure 4.31, two half-wavelength dipole elements are placed in the axis of a semi-cylindrical reflector with radius $R = 60$ mm. The length of the reflector is $H = 250$ mm and the inter-element spacing D_e is set to be 60 mm, which is about half of the free-space wavelength at 2.45 GHz. The simplest method to excite a uniform linear array is to excite the array elements with equal amplitude and a progressive phase shift $\Delta\psi$ [31]. For base station applications, the maximum phase shift is given by the number of array elements N and the scan angle θ [32]

$$\Delta\varphi_{\max} = -(N-1) \times k \times d \times \sin\theta \tag{4.18}$$

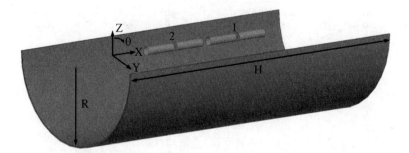

Figure 4.31 Schematic view of the two-element fed metallic reflector presented in [28].

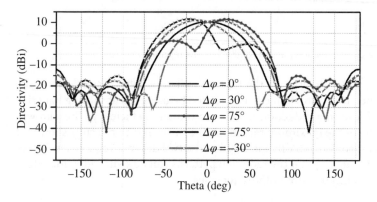

Figure 4.32 Beam scan patterns of the two-element fed metallic reflector.

where $k = 2\pi/\lambda$ and $d = 0.5\lambda$. Figure 4.32 shows that the achievable beam scan range of the two-element fed metallic reflector is up to ±25°, with the maximum sidelobe level (SLL) below −10 dB. The simulated directivity variation within the beam scan range is less than 2 dB. To optimise the beam-tilting performance, three parameters, R, H, and D_e, are studied. The length of the reflector has almost no effect on the directivity, while for the shorter reflectors, the SLL is increased by 5 dB. Larger R leads to extended beam tilt angle at the cost of increased SLL. Furthermore, enlarging the spacing between elements slightly degrades the SLL. Based on the above studies, the dimensions of the reflector and the element spacing can be optimised and used for the active FSS reflector design.

After the preliminary evaluation of the array-fed semi-cylindrical reflector antenna, the key to switching radiation pattern in the horizontal plane is to design a reconfigurable FSS cylinder that is capable of reconfiguring the omnidirectional beam of the feed antenna to a directional one. An H-shaped AFSS unit cell based on a switchable slot is designed to operate at 2.45 GHz, as shown in Figure 4.33. Similar to a dipole antenna, a slot FSS resonates when the length of the slot is one half-wavelength. According to Babinet's principle [33], the slot FSS with slot length S_x has a resonance frequency around

$$f_c = \frac{1}{2S_x\sqrt{\varepsilon\mu}} \tag{4.19}$$

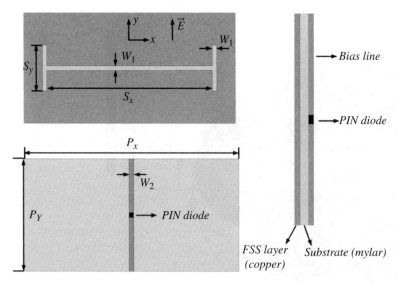

Figure 4.33 Schematic view of the AFSS unit cell reported in [28].

More specifically, with slot width S_y, the resonance frequency can be calculated by

$$f_c = \frac{C}{2.1\left(1 + \dfrac{S_y}{2S_x}\right)S_x} \tag{4.20}$$

A thin polyester film with $\varepsilon_r = 2.7$ and loss tangent of 0.0023 is chosen as the substrate because it is flexible, durable, and suitable for conformal antenna design. The dimensions of the unit cell are given in Table 4.1. The whole structure is simulated in HFSS where the PIN diode states are modeled as series or parallel RLC lumped elements. For the ON state, the diode is simplified as a forward resistor $R_s = 2.1\ \Omega$, and as an equivalent capacitor $C_s = 0.17$ pF in the OFF state. Note that the H-shaped slot reduces the FSS array periodicity P_x, which can be beneficial, enabling accommodation of more cells for a fixed cylinder perimeter.

Based on the previous discussion, the planar AFSS is rolled into a cylinder which has the same dimensions as the metallic reflector. First, the numbers of unit cells along the circumference and axis (N_x and N_y) are calculated from

$$N_x = 2\pi R/P_x \tag{4.21}$$

$$N_y = L/P_y \tag{4.22}$$

where $L = 190$ mm. Then, same antenna arrays used for the metallic reflector are employed to illuminate the AFSS reflector. Note that the more unit cells there are along

Table 4.1 Dimensions of the slot FSS unit cell (mm).

S_x	S_y	W_1	W_2	P_x	P_y
37	10	1	1	47	23.75

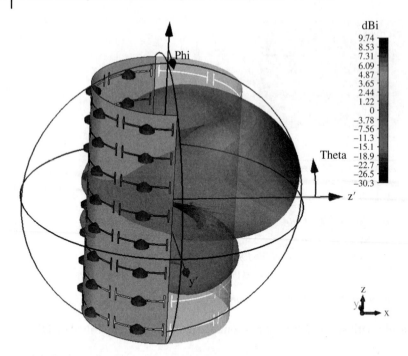

Figure 4.34 CST radiation pattern simulation of AFSS uptilted beam state fed by the two-dipole array.

the cylinder circumference, the more beams that can be generated in the horizontal plane. Figure 4.34 shows the simulated radiation pattern of the AFSS cylinder fed by the two-element dipole array when $\Delta\psi$ is set to 90°. The whole cylinder consists of eight columns and eight rows of unit cells with the dimensions given in Table 4.1. The PIN diodes in the left-half cylinder are switched ON to activate the reflecting mode whereas the other half diodes are switched OFF, which allows the EM wave to pass through the right-half cylinder (which is made transparent for figure clarity). By varying the phase shift between the two ports, similar beam-tilting trends can be obtained, as shown in Figure 4.35. Compared with Figure 4.32, the maximum beam tilt angle with an SLL of −10 dB is reduced to ±21°. Accordingly, a four-element fed AFSS cylinder is designed using the same setup as for the metallic counterpart. As seen in Figure 4.36, a more directional beam can be generated with a maximum beam scan angle of ±30°.

It is required that the dipole array must be placed along the axis of the AFSS cylinder to maintain the radiation pattern symmetry when performing the beam-switching. Practical dipole array designs for base stations can be found in [34], where conventional corporate feed networks can be hidden behind or near a metal ground plane to minimise the effect on the radiation pattern of the antenna. Nevertheless, for this design, implementing an antenna array involves designing a power-combining network for the radiating elements, which leads to antenna structure asymmetry and performance deterioration when placed inside the AFSS cylinder due to multiple reflections and strong coupling inside the cylinder. To address this problem, a multi-layer omnidirectional antenna array can be used to overcome the disadvantages of the conventional collinear dipole array.

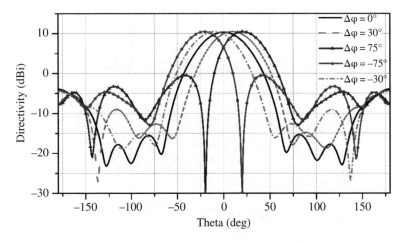

Figure 4.35 Simulated AFSS antenna [28] uptilt radiation patterns fed by the two-dipole array in the elevation plane.

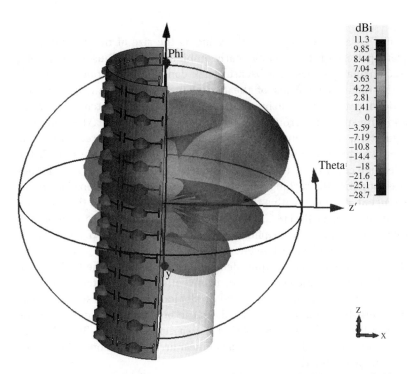

Figure 4.36 CST radiation pattern simulation of the AFSS uptilted beam state fed by a four-dipole array.

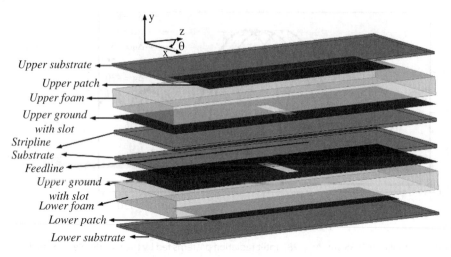

Figure 4.37 Details of the multi-layer array element design presented in [28].

As described in [35] and [36], significantly reducing the size of the ground plane makes the H-plane pattern of the antenna nearly omnidirectional. Thus, by arranging two identical rectangular patches in a back-to-back configuration along the broadside direction, the radiated far-fields of the two elements are summed to form a near-omnidirectional pattern in the azimuth plane. A back-to-back patch antenna fed by a stripline and with small size ground planes is designed to feed the AFSS antenna. With this configuration, two rectangular patches radiate in opposite directions in the horizontal plane to form an omnidirectional pattern. As shown in Figure 4.37, the multi-layered element consists of two identical patches which are fed by striplines through aperture coupling. The rectangular patches are etched on an Arlon Cuclad 217 ($\varepsilon_r = 2.17$, loss tangent = 0.0009) with a thickness of 0.508 mm. Because the substrate is quite flexible, two pieces of ROHACELL foam are used as spacers to support the patches. The substrate used for the stripline is 0.762 mm thick Arlon AD255A with a relative permittivity of 2.55 and a loss tangent of 0.0015.

Figure 4.38 shows the simulated reflection coefficient of the antenna. The 10 dB return loss bandwidth is about 4.7% ,which is similar to a conventional patch antenna. In order to tune the input impedance, the slot sizes L_a and L_b, slot location L_2, and stub length L_1 are critical parameters that should be optimised. As can be seen from Figure 4.39, the radiation patterns in the E and H planes are similar to a dipole antenna. Thus, by tuning the phases of several identical elements in a collinear array, a tilted omnidirectional radiation pattern can be obtained.

Using the element shown in Figure 4.37, a two-element parallel fed array is designed and measured. Figure 4.40 shows the configuration of the array antenna. The feeding points at the bottom can be connected to phase shifters and a feed network through a stripline to coaxial line transition. To avoid grating lobes, the element spacing d_e is kept around 0.5–0.7λ at the desired frequency. To maintain good isolation between the two elements, the feedline of the upper element is offset by a distance from the coupling slot of the lower element. The stripline is designed to have a 50 Ω characteristic impedance.

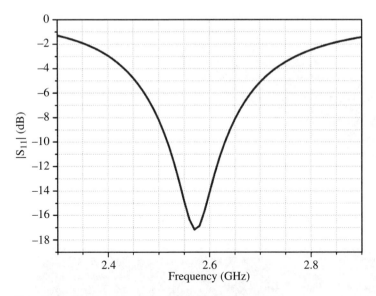

Figure 4.38 Simulated S_{11} of the back-to-back patch antenna.

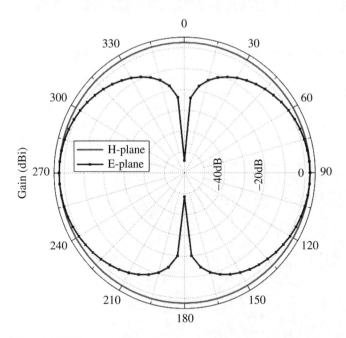

Figure 4.39 Simulated radiation patterns of the back-to-back patch antenna.

One advantage of this design over the conventional metallic collinear dipole array is that the feedlines are hidden between the grounds, which avoids interference from the feed lines to the radiating elements. As shown in Figure 4.41, there is negligible induced current on the bottom element when the upper element is excited, which indicates that the isolation between them is high.

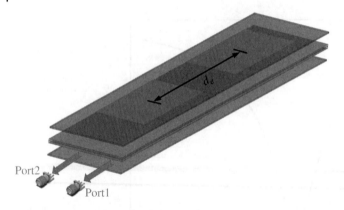

Figure 4.40 Perspective view of the multi-layer antenna feed array reported in [28].

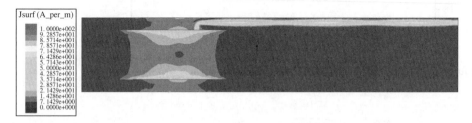

Figure 4.41 Simulated current distribution at 2.45 GHz when the upper element is excited.

Before integrating the AFSS and the feed array, measurements were performed to confirm the performance of the AFSS. First, the transmission coefficients of the planar AFSS were measured in a plane wave chamber. The planar FSS comprised 8 × 8 unit cells. As can be seen in Figure 4.42, eight parallel lines with PIN diodes in series are etched on one side of the substrate, whereas on the other side is the copper layer with FSS slots, which

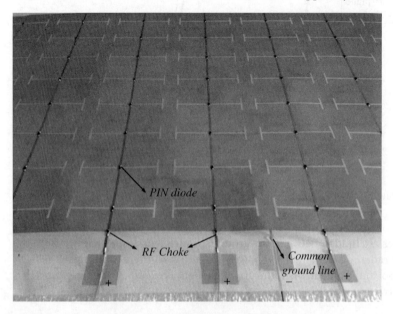

Figure 4.42 Fabricated bias network of the AFSS reported in [28].

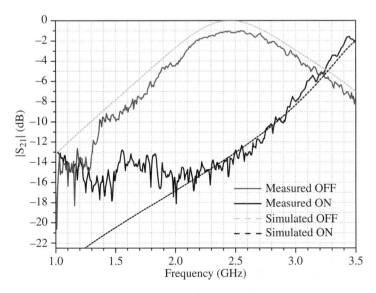

Figure 4.43 Measured transmission coefficients of the AFSS reported in [28].

is used as the common ground for the bias lines through a metallic via. Thus, it is convenient to control each column with a 5 V DC voltage, and there is only one line connected to the ground. RF inductors are used as RF chokes to isolate the RF signal, and a 1 KΩ resistor is added to the bias line to limit the current. Figure 4.43 compares the simulated and measured values of transmission coefficients when switching ON/OFF all the PIN diodes. It can be noted that the performance of the fabricated planar AFSSs agrees well with the simulated ones except that the prototype shows 1 dB higher insertion loss.

The fabricated antenna array is shown in Figure 4.44. Figure 4.45 shows the measured *S* parameters. The measured port isolation is less than the predicted results but still larger than 20 dB, which is sufficient for most applications. Two coaxial cables are connected

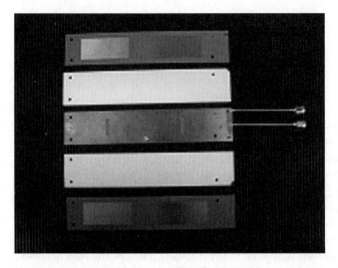

Figure 4.44 Fabricated feed array [28] before assembly.

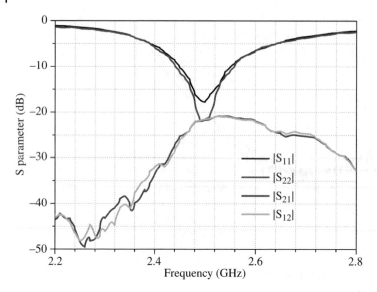

Figure 4.45 Measured *S* parameters of the fabricated array reported in [28].

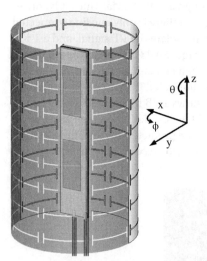

Figure 4.46 Perspective view of the AFSS antenna CST model.

to the feed ports of the two antenna elements, which is the interface to connect the external power-combining circuit with the required phase shifts.

Figure 4.46 shows the simulation model of the AFSS beam-switching antenna. Note that all the PIN diodes in the gaps of the slots are not shown for figure clarity. The feed array is placed in the centre of the cylinder parallel to the *x*–*z* plane. The feed points are extended beyond the cylinder to connect the feed network. With this configuration, the solid sector acts as a reflector by switching on all the diodes on it. The phase shifts required are obtained by optimising the progressive phase difference and element spacing. Then the required phase shift is applied to the microstrip line of the feed network, allowing the array pattern to be tilted to the desired angle.

Figure 4.47 Fabricated entire AFSS antenna under test [28].

The fabricated antenna prototype is illustrated in Figure 4.47. The antenna has impedance bandwidth 2.4–2.5 GHz. Meanwhile, the isolation of the two ports is found to be reduced to around 14 dB due to the presence of the AFSS cylinder. It should be noted that when the feed array is placed in the centre of the cylinder, the isolation of the antenna ports is degraded. To compensate for this effect, the coupling between elements and the element reflection coefficient have to be kept as small as possible. The array element spacing is a critical parameter to tune the coupling, but it also affects the beam scan angle and SLL. According to the simulation results, in this design the required phase shifts for the up- and downtilt beams are 45° and 245°, respectively, when the beams are pointed at ±15° with maximum −10 dB SLLs. Thus, two individual feed network circuits, including the required phase shift values, are simulated with the structure. Figure 4.48 compares the simulated and measured reflection coefficients of the final

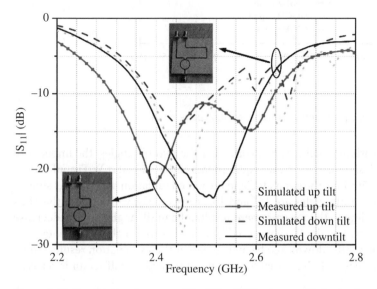

Figure 4.48 Simulated and measured S_{11} of the AFSS antenna with feed network [28].

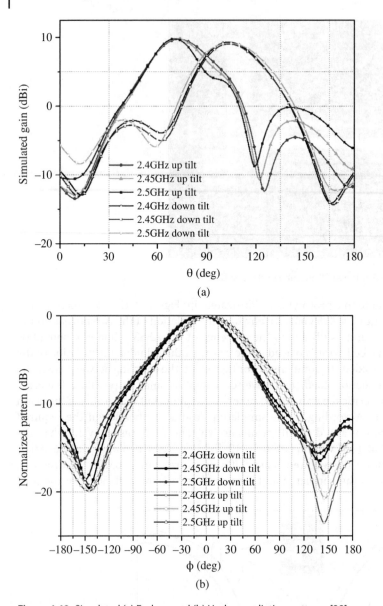

Figure 4.49 Simulated (a) E-plane and (b) H-plane radiation patterns [28].

antenna structure with external feed networks. For the two tilt angles, the impedance bandwidth of both designs can meet the requirements of the 2.45 GHz WiFi application.

Figures 4.49 and 4.50 show the simulated and measured radiation patterns. As shown, the up and down beam tilt angles are 16° and 15°, respectively, with both of the SLLs below −10dB. According to the measurement results, the uptilt angle has a 2° decrease at the lower frequency, but the SLL has a 3 dB increase at the higher frequency. The measured scan angle for the downtilted beam is 2° larger than simulation results. There is also a slight increase in SLL. Note that for the measurements a half sector is selected

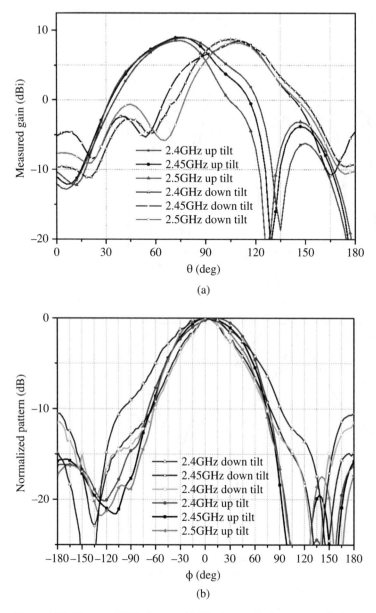

Figure 4.50 Measured (a) E-plane and (b) H-plane radiation patterns [28].

and biased to act as a reflector. Eight identical up/down beams can be generated if half of the AFSS cylinder is switched ON/OFF in turn because of the symmetry of the antenna structure. Figure 4.51 shows the measured eight beams in the horizontal plane for the up- and downtilt angles at 2.45 GHz. Compared with the simulated results, the scan angle for the same SLL (−10 dB) is reduced mainly because the AFSS is not an ideal reflector. Furthermore, unwanted scattering happens when EM waves pass through the OFF-state AFSS, which degrades the beam-tilting performance as well.

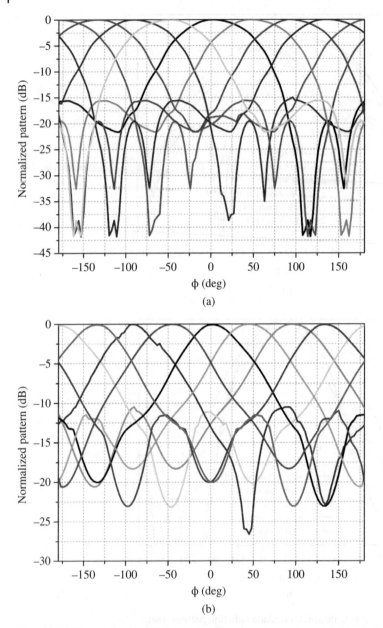

Figure 4.51 Measured (a) uptilt and (b) downtilt radiation pattern switching in the horizontal plane.

4.6 Frequency-agile Beam-switchable Antenna

Loading FSSs with varactors introduces more degrees of design freedom compared with PIN diodes [37–40]. This section introduces the FSS-based frequency-tuning antenna reported in [41]. As a preliminary examination of the tunability of the slot FSS, a single layer model is simulated in the full-wave EM solver CST. The reconfigurable FSS

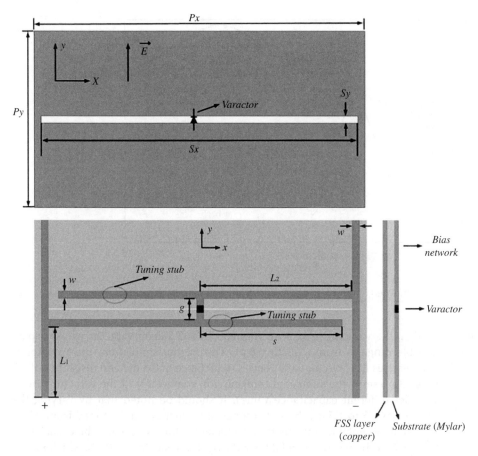

Figure 4.52 Schematic views of the slot FSS unit cell.

unit cell consists of a half-wavelength slot in the center of a copper plate, as shown in Figure 4.52. A varactor is connected in the gap, and the bias line is excluded for simplicity. The dimensions of the unit cell are $P_x = 48$ mm, $P_y = 25$ mm, $S_x = 46$ mm, and $S_y = 0.2$ mm. The slot FSS is a relatively simple structure which makes it a good option for AFSS design as it can lower the complexity of the biasing circuit. To better understand the tuning mechanism of the FSS, an equivalent circuit model has been developed. The equivalent circuit model is a parallel L–C circuit. The inductance is associated with the electric current flowing in the patch, and the capacitance consists of an intrinsic capacitance C between the gap of the slot and the variable capacitance C_v from the varactor diode. According to bandpass filter theory, the resonance frequency of the FSS can be expressed by

$$f = \frac{1}{2\pi\sqrt{L(C + C_v)}} \tag{4.23}$$

From this equation it can be seen that increasing the capacitance of the varactor decreases the resonance frequency. The tuning range is limited by the capacitance ratio of the varactor. Higher ratios normally yield wider frequency-tuning ranges.

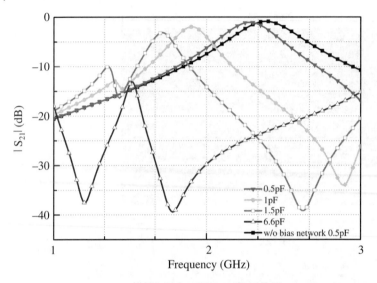

Figure 4.53 Simulated transmission coefficients of the unit cell with varactor of different capacitance values.

In this design, the varactor used is an Infineon BB857 silicon varactor whose capacitance tunable range is from 0.52 to 6.6 pF. During the simulation, the varactor is modeled as a series RLC circuit with a series inductance of 0.6 nH and resistance of 1.5 Ω. Figure 4.53 presents the simulated transmission coefficients of the varactor-loaded slot FSS under different capacitance values. It should be noted that by varying the capacitance from 0.5 to 1.5 pF, the slot resonance frequency can be tuned from 2.5 to 1.7 GHz with an acceptable transmission coefficient. Moreover, when the capacitance increased to 6.6 pF, between 1.2 and 3 GHz the S_{21} level is below -20 dB, which means almost no EM wave can be transmitted through the FSS. The high reflection level at 6.6 pF can be used for reconfiguring the FSS to reflect the incoming waves.

The bias network is one of the most critical parts of the active FSS. Maintaining a simple network is beneficial for minimising the interferences from the bias lines [42]. In the design, it is preferred that all the varactors be parallel fed to limit the maximum control voltage. The substrate is a 0.05 mm thick Mylar film with $\varepsilon_r = 2.7$ and loss tangent of 0.0023. Two vertical DC main control lines are located at each side of the slot, providing the positive and negative biasing for the varactor. The package size of the diode is about 2×1 mm so the gap between the two pads for bridging the varactor and the width of the bias line is set to be 1 mm. The dimensions of the bias network are given in Table 4.2.

As an indispensable part of the ASS design, the bias network needs to be well designed in order to minimise its effect on the frequency response of the FSS. In some cases

Table 4.2 Dimensions of the bias circuit (mm)

L_1	L_2	w	g	s	P_x	P_y
10.5	22.5	1	2	20.5	48	25

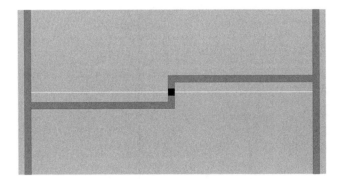

Figure 4.54 Initial design of the parallel bias network.

even the bias lines are printed on the different layers, the frequency response of the FSS deteriorates when the network is complex and has a size that is comparable to the wavelength. In this case, the overall structure can be treated as a bandpass FSS layer (i.e. the slots) cascading a bandstop patch-type FSS (i.e. the bias lines), as the incident wave excites the bias network, especially when the AFSS is close to an antenna feed. Thus, as shown in Figure 4.52, two tuning stubs are used to maintain the desired transmission responses. Figure 4.54 shows the layout of a simple bias network without the two tuning stubs. The simulated current distributions at 2.4 GHz for the two cases are presented in Figure 4.55. It is clear that without the stubs most of the induced currents are concentrated on the bias lines close to the slot. The unwanted coupling deteriorates

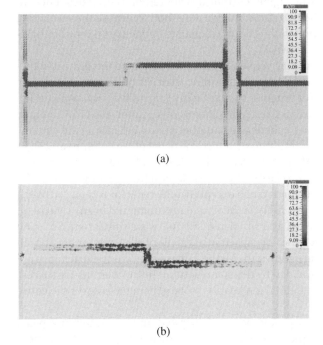

Figure 4.55 Simulated current distributions on the bias network at 2.4 GHz: (a) original case and (b) with stubs.

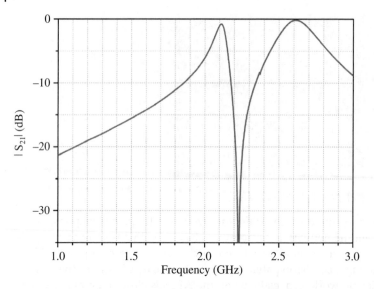

Figure 4.56 Simulated transmission coefficient of the bias network without stubs.

the transmission coefficients, as shown in Figure 4.56. There is a bandstop pole from 2.2 to 2.4 GHz, which is exactly at the desired working frequencies. It can be noted from Figure 4.55b that with the tuning stubs the induced currents concentrate on the half-tuning stubs only. There is negligible current distributed on the vertical lines. In light of the discussion above, the two stubs provide a method to detune the resonating bias network, enabling a reduction in the circuit resonance frequency and thus mitigating the effect caused by the bias network.

After the investigation of the planar active FSS, the next step is to integrate it with a feed. The planar FSS is rolled into a cylinder to mimic a corner reflector antenna. It is necessary that a metallic reflector antenna should be investigated to give some design guidance about the influence of AFSS size, feeder dimensions, etc. for the AFSS antenna. Before rolling the planar FSS into a cylinder, the number of unit cells along the circumference and axis is calculated. The resulting structure consisted of ten columns and eight rows, as shown in Figure 4.57. Note that a half-cylindrical AFSS and diodes are made transparent for the purpose of figure clarity. The dimensions of the AFSS antenna are the same as those of the metallic reflector antenna except that the reflector is replaced by the cylindrical AFSS. To explore the feasibility of the frequency tuning and beam-switching, the entire configuration is modelled and simulated in a full-wave simulator CST using the frequency-domain solver. A series of capacitance values of the varactor are chosen in order to justify the operational bandwidth. All the values of the diode capacitance are extracted from the data sheet provided by the manufacturer. If five columns are successively reconfigured as a reflector, in total ten beams can be generated to cover the entire horizontal plane.

To verify the design concept, a prototype was fabricated and measured in a plane wave chamber. A photograph of the AFSS under test is shown in Figure 4.58. The measured transmission response of the tunable FSS at different DC voltages is presented in Figure 4.59. As can be seen from this figure, the resonance frequency increased from

Figure 4.57 CST model of the resulting antenna
structure reported in [41].

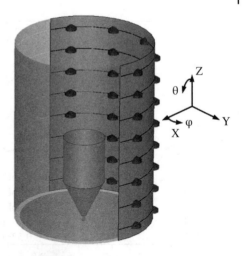

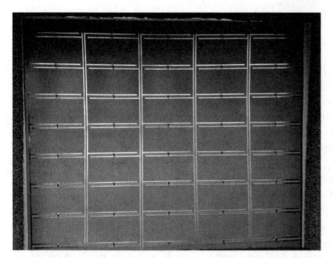

Figure 4.58 Fabricated bias circuit on one side of the substrate [41].

1.75 GHz on 7.6 V to 2.35 GHz on 28.1 V, while the bandpass insertion loss is reduced
from 3 to 0.5 dB. Moreover, when the capacitance is tuned to 6.6 pF at 0 V, the FSS is
switched to reflection mode as the transmission coefficients are all below −12 dB across
the tuning range. This feature is useful to the design of low-cost smart antennas because
in this state the AFSS antenna requires zero control power.

The prototype of the AFSS antenna was fabricated and a polystyrene foam used to support the FSS cylinder. The DC control lines are hidden under the metal ground plane of
the monocone antenna, shielding the unwanted radiation from the cables. To implement
the biasing, half of the AFSS cylinder is required to be applied with a variable voltage
and the other half with zero bias voltage. One advantage of the bias network design is
the simplified the DC control system: all the positive lines are combined and connected
to the positive of the power supply, which has a tuning range of 0–30 V, hence switching
ON/OFF the DC negative lines can control each column of the AFSS cylinder. Moreover,

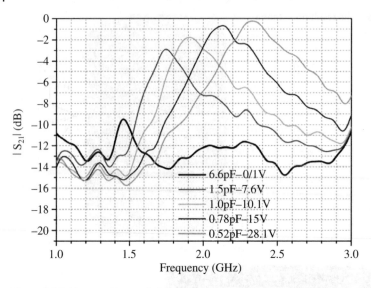

Figure 4.59 Measured transmission coefficients of the tunable FSS [41].

to switch ON the diodes on half of the cylinder, five DC negative lines are connected to the negative of the power supply. When scaling this design into a larger size array, the positive voltage bias lines become the common ground while the structure can be controlled through varying negative voltages. Figure 4.60 compares the measured reflection coefficients versus frequency at the corresponding bias voltages. It should be noted that the antenna impedance matching band can be tuned with the applied voltage. Theoretically, the poles of the S_{11} are in accordance with the resonance frequencies of the AFSS transmission coefficient. Note that there is a slight frequency shift at most of the bias voltages and a 150 MHz discrepancy at 28.1 V. However, the antenna remains well matched at each range of the operating frequency.

Figure 4.61 shows the measured gains at the four bias voltages. The peak gain is tuned from 1.7 to 2.3 GHz, which corresponds to a 30% frequency range. It can be seen that the gain tuning range is close to the measured AFSS bandpass shown in Figure 4.59. Since the FSS can be regarded as a spatial filter, when it is integrated with an antenna the gain tunability of the resulting antenna follows the same trend of bandpass tunability of the filter.

Radiation patterns were measured with several diode control voltages. During the measurement, half of the AFSS cylinder (five columns) is forward biased while the other half is connected to zero bias voltage. The same antenna performance will be realised if different half AFSS cylinders are selected owing to the structural symmetry. Figure 4.62 shows the measured results in the E and H planes at the peak gain frequencies. For most cases, there is good agreement between the simulation and measurement. The higher level side lobe and back lobe found in the two planes are the consequence of the imperfect replacement of the metallic reflector. It can be seen that for the E-plane radiation pattern at 1.7 GHz, the side lobe is at −9.5 dB. Figure 4.63 shows the measured switched beams in the H plane at 1.9 GHz. It is demonstrated that by successively selecting half of the AFSS cylinder, the generated beams are capable of covering the entire horizontal plane.

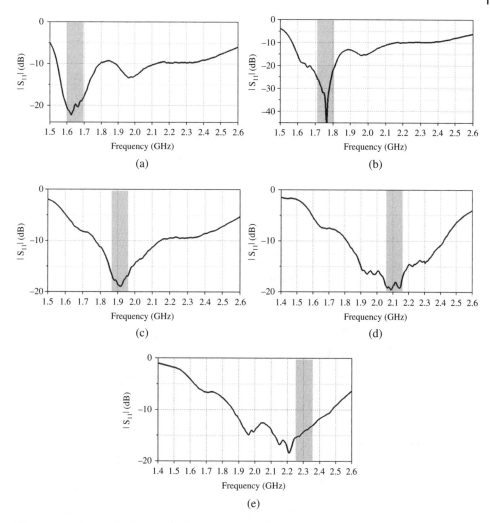

Figure 4.60 Measured reflection coefficients at (a) 7 V, (b) 8 V (c) 10.1 V, (d) 15 V, and (e) 28.1 V [41].

4.7 Continuous Beam-steering Antennas using FSSs

The antennas discussed in the previous sections are based on the concept of corner reflector antenna. Using the cylindrical active FSS, several predefined beams can be steered at discrete angles in the azimuth plane. In some applications, it is required that the beam of the antenna can be continuous steered. Thus, in this section some techniques utilising FSSs to realise continuous beam steering antennas are presented.

In [43], a continuous beam steerable planar antenna using a varactor-loaded FSS is reported. The antenna consists of a wideband bow-tie radiating element mounted above an active artificial magnetic conductor (AMC). The AMC consists of an FSS printed on a 1.52 mm thick Taconic RF60 grounded dielectric substrate. As shown in Figure 4.64, a 5.4 × 5.4 mm square metallic patch is used as the unit cell. Varactors are placed between

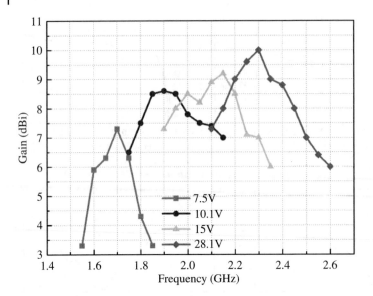

Figure 4.61 Measured gain tuning range for the four bias voltages [41].

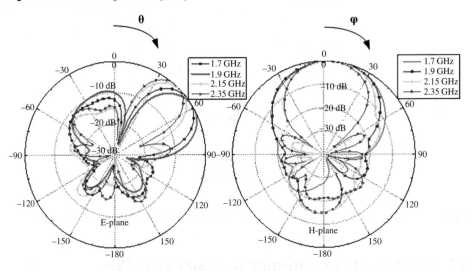

Figure 4.62 Measured radiation patterns at the peak gain frequencies [41].

each patch element and the ground plane through a metalised via. With this parallel biasing scheme, the varactors can be simply fed by applying a positive voltage to the FSS and a negative voltage to the ground plane. The 5 × 5 patch grid is arranged into five rows, and each element is electrically connected by narrow striplines with a resistor of 10 kΩ at their ends. The Abrupt Tuning Varactor MTV-30-05-08 is soldered to the FSS by wires bonding. The series resistance takes into account the diode resistance and the electrical contact. A high-Q varactor can keep the series resistance down to 0.3 Ω. An RLC model of the varactor is employed to tune the diode capacitance. The simulated AMC reflection phase is shown in Figure 4.65. The reflection coefficient is calculated at normal incidence with the electric field parallel to the diode orientation.

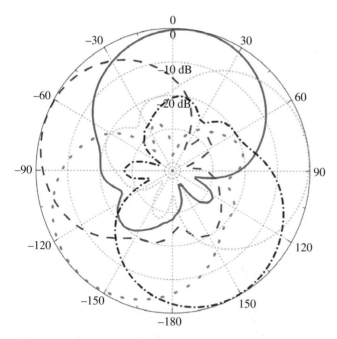

Figure 4.63 Measured beam-switching characteristic at 1.9 GHz in the H plane [41].

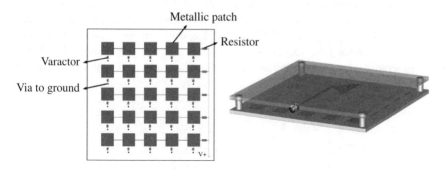

Figure 4.64 Schematic views of the tunable AMC and resulting antenna reported in [43].

The feed antenna is printed on both sides of a 0.8 mm thick FR4 substrate and fed by two 50 Ω parallel stripes. The operating band of the bow-tie is designed to match the tuning range of the active AMC surface. The radiation properties of the proposed structure can be explained by using leaky wave theories. Radiation or coupling between the space wave and the surface wave requires that the wave vector of the space wave k_0 must have a component tangential to the surface that matches the wave vector of the surface wave k_y. Radiation cannot occur when $\omega < ck_y$ as phase matching cannot occur for any angle, thus the wave is bound to the surface. When $\omega > ck_y$, the EM wave can radiate from the surface into free space, at an angle given by

$$\theta = \sin^{-1}\left(\frac{ck_y}{\omega}\right) \tag{4.24}$$

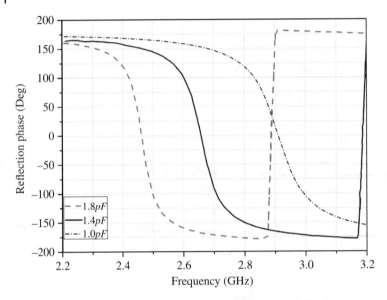

Figure 4.65 Phase response of the AMC reported in [43] with different varactor diode capacitances.

More detailed explanation of this phenomena can be found in [44] and [45]. Beam-scanning towards positive and negative angles with respect to the normal direction can be achieved due to the generation of forward and backward leaky wave radiation. Figure 4.66 shows the simulated radiation patterns when different capacitance values are applied at 2.76 GHz. By varying the varactor capacitance around 1 pF, a scan range of about ±30° can be obtained. Note that the gain is plotted in linear scale. This reported design shows that surface waves of an active FSS can be controlled to realise an electronically steerable antenna.

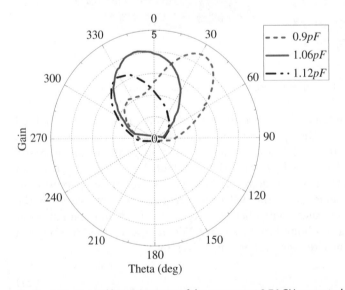

Figure 4.66 Steered beam patterns of the antenna at 2.76 GHz reported in [43].

Figure 4.67 Configuration of the FSS unit cell reported in [46].

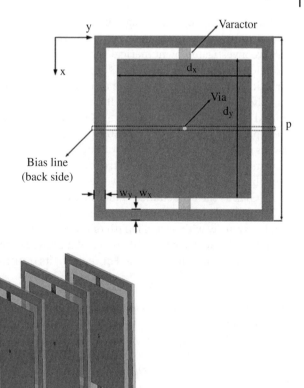

Figure 4.68 Configuration of the five-layer FSS reported in [46].

An AFSS-based high-gain beam-steering antenna is presented in [46], where the FSS is tuned to obtain the gradient phase distribution of the emitted wave along the E and H planes. By placing the varactor controlled FSS in front of a horn antenna, continuous beam-steering in both planes can be achieved. Figure 4.67 shows the unit cell of the FSS. The bandpass FSS consists of a metallic rectangular loop slot etched on one side of a 1.524 mm thick Taconic TLX-8 substrate. The relative permittivities of the material and loss tangent are 2.55 and 0.0019, respectively. A pair of varactor diodes is soldered to the center of the upper and bottom ring slots along the y direction. To tune the diodes, bias lines are etched on the other side of the substrate, and the passive FSS element is connected to the bias circuit through a metallic via. At its resonance, the rectangular ring along the x direction yields an equivalent capacitance C_e, while the effective inductance from the metallic strips of the rectangular ring along the y direction is L. A SKYWORKS SMV1405- 079LF varactor diode was chosen for this design and its effective capacitance is C_v. The resonant frequency of the active FSS can be calculated by

$$\omega_0 = \frac{2}{\sqrt{L(C_v + C_e)}} \tag{4.25}$$

Thus, by applying DC voltage to the diodes, the resonant frequency can be continuously tuned within a frequency range. The transmission responses of the five-layer FSS are investigated by numerical simulation with an incident plane wave along the $+z$ direction, as shown in Figure 4.68. The electric field of the incident wave is along the y direction. Figure 4.69a shows the transmission coefficient of the infinite five-layer FSS for different capacitance values. It can be seen that a wideband bandpass response is shifted from 5.2–5.7 GHz to 4.6–5.4 GHz with a maximum insertion loss of 3 dB when the capacitance varies from 2.4 to 0.8 pF. Note that the operating frequency of 5.3 GHz is always in the bandpass for all the tuning states, so most of the electromagnetic wave can transmit through the five-layer structure at 5.3 GHz. Figure 4.69b shows the transmission phase when the capacitance varies. As shown, a 360° phase change can be obtained at 5.3 GHz as the capacitance value varies from 0.65 to 2.6 pF [47, 48].

The principle of the antenna beam-steering using the multilayer FSS is similar to that of a linear array with a total length of l. Divided into N regions, the linear arrays have a period of l/N, with a phase difference of α between adjacent regions. The steered beam angle θ is then determined by α. For region i, its transmission phase ϕ_i can be defined as

$$\varphi_i = -\varsigma_i + \varphi_0 + 2k\pi \tag{4.26}$$

where φ_0 is a constant and k is an integer. The phase shift ς_i and phase difference α can be described as

$$\varsigma_i = i\beta l \sin\theta/N \tag{4.27}$$

$$\alpha = \varphi_i - \varphi_{i-1} = -\beta l \sin\theta/N \tag{4.28}$$

where β is the wave number. Using Equations (4.26), (4.27), and (4.28), the beam-steering angle θ can be derived

$$\theta = \arcsin\left(-\frac{N\alpha}{\beta l}\right) \tag{4.29}$$

In order to demonstrate the beam-steering performance, a horn antenna with an aperture size of 184×206 mm is used as the feed antenna. The five-layered FSS is placed in front of the horn aperture, and the air gap between them is optimised to be 18.5 mm. The horn aperture is close to the FSS so the active FSS re-radiates the incident quasi-plane wave with the E field in the y direction. The schematics of the whole beam-steering antenna design are shown in Figure 4.70. The active FSS consists of five identical layers and each layer consists of 6×6 unit cells. To validate the beam-steering performance of the antenna, four different beam angles, 0°, 10°, 20°, and 30°, are selected, and the corresponding phase differences are calculated. According to (4.28) and (4.29), the phase difference between adjacent columns/rows α can be calculated as 0°, 36°, 70°, and 103°. For the beam-steering in the E plane, the same voltage Φ_{Hn} ($n = 1, 2...6$) should be applied to the square ring, while the phase variation is tuned by the control voltage Φ_{Em}. The simulated S_{11} of the resulted antenna for each scan angle is shown in Figure 4.71a. At 5.3 GHz, the S_{11} of the four cases are all below -10 dB. Figure 4.71b shows the simulated radiation patterns. The steered main lobe angles are 0°, 11°, 21°, and 29.5°, respectively. As the antenna scans from 0° to 29.5°, the simulated gain decreases from 18.2 to 16.6 dBi. With this configuration, beam-steering can also be achieved in the H plane. The DC control voltage Φ_{Em} ($m = 1, 2...6$) applied to the bias lines is the same while Φ_{Hn} ($n = 1, 2...6$) applied to the metal rings are different in order to obtain the required

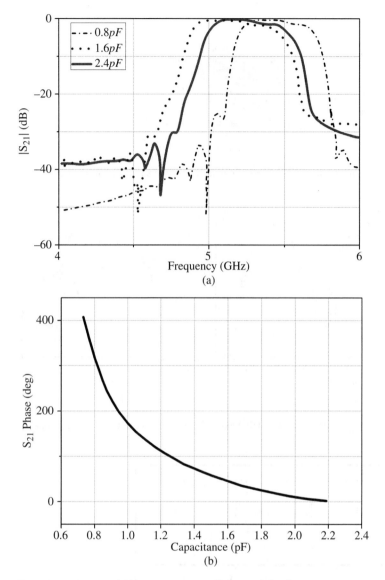

Figure 4.69 Measured (a) transmission coefficient and (b) phase variation range of the five-layer FSS reported in [46].

phase distribution across the entire surface. The beam-steering in the H plane has similar performance to that of the E plane because of the symmetrical structure.

The horn antenna incorporating the active FSS was fabricated and characterised. First, the transmission amplitude and phase of the active FSS are measured. Figure 4.72 shows the measured S_{21} and its phase for different configurations of the bias voltages applied on the varactor diodes. It can be noted that when the bias voltage is changed from 0 to 30 V, the transmission phase is almost linearly varied from 0° to 427°. The insertion loss is mainly caused by the diode resistive loss, substrate loss, and the

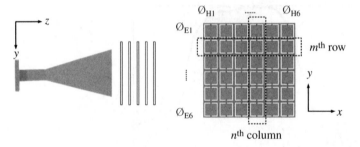

Figure 4.70 Horn antenna incorporating AFSS and schematic of the AFSS phase control described in [46].

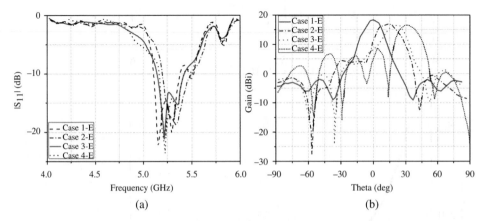

Figure 4.71 (a) Simulated S_{11} of the antenna and (b) steered beam patterns for the four scan angles reported in [46].

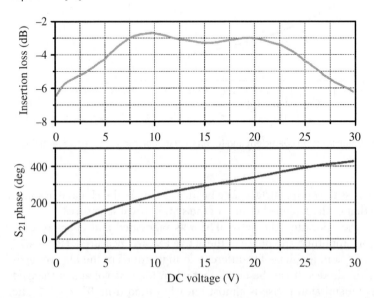

Figure 4.72 Measured insertion loss and S_{21} at different DC control voltages of the AFSS reported in [46].

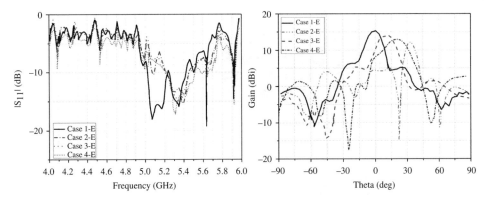

Figure 4.73 Measured S_{11} and steered beam patterns for the four scanning angles of the antenna presented in [46].

reflection between layers. The maximum insertion loss is 6.5 dB at zero voltage. For E-plane beam-scanning, the voltage Φ_{Hn} ($n = 1, 2...6$) is set to be 30 V and different values are set for Φ_{Em} ($m = 1, 2...6$) so the bias voltage of the varactor diodes in the row can be expressed as $V_m = \Phi_{Hn} - \Phi_{Em}$. The broadside radiation can be obtained by applying the same bias voltages to all rows. Figure 4.73 shows the measured results of the antenna. The S_{11} are less than -10 dB at 5.25–5.43 GHz for all the four scan angles. Integrating the active FSS to achieve beam-steering leads to larger than 3.5 dB antenna loss, mostly from the loss of FSS.

4.8 Case Study

In the previous sections different types of beam steering antennas using active FSSs are presented. This section aims to provide a step-by-step guide to designing a beam-switching antenna based on slot-active FSS. Design procedures and full-wave simulation results are elaborated so that the reader can follow the guidelines to complete the antenna design. In this case study, a low-cost electronic beam-switching antenna operating at the 2.4 GHz band is presented. In comparison with the design in [26], the dimensions of the antenna in this study are reduced to achieve a more compact size. Low-cost PIN diodes are used to design this beam-switching antenna, which consists of two parts: an active FSS cylinder and a feed antenna. The antenna operating principle is introduced in section 4.3. A sleeve monopole antenna is chosen as the feed, as shown in Figure 4.74. The antenna impedance matching is tuned by adjusting the sleeve height L_c, monopole height L_m, and diameter of the radiator D_m. The dimensions of the feed antenna are optimised to have a wideband impedance bandwidth. Here we choose $D_m = 2$ mm, $L_c = 0$ mm, $L_m = 30$ mm, and $R_g = 61$ mm.

After selecting the feed antenna, the next step is the active slot FSS design. The aim is to load PIN diodes into the FSS slots to reflect/transmit the incident wave by switching the diodes ON/OFF. As validated in previous sections, integrating a PIN diode into the FSS decreases the resonant frequency of the FSS as a result of the added capacitance to the slot. Thus, before loading the PIN diode, a passive FSS with a resonant frequency higher than the operating frequency should be designed and simulated. Figure 4.75 shows the

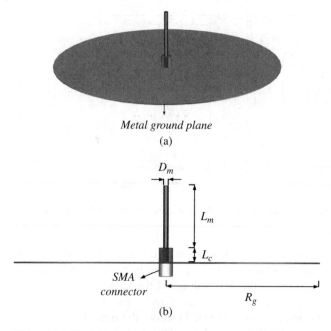

Metal ground plane

(a)

D_m

L_m

L_c

SMA
connector

R_g

(b)

Figure 4.74 Sleeve monopole antenna: (a) 3D view and (b) side view.

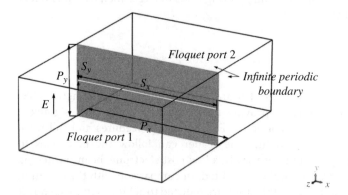

Floquet port 2

S_y

P_y

S_x

Infinite periodic
boundary

E

P_x

Floquet port 1

Figure 4.75 FSS unit cell model in CST.

CST model of the passive slot FSS. It was modelled using the Floquet mode settings of the frequency domain solver [49]. The bandpass resonant frequency is mainly tuned by P_X and S_X, while S_Y and P_Y have an impact on the bandwidth of the bandpass. As a starting point, the resonant frequency f_0 is set to 2.8 GHz. As a rule of thumb, the slot length should be approximately half-wavelength at the resonant frequency. Here we choose $P_X = 55$ mm, $S_X = 54.5$ mm, $S_y = 0.5$ mm, and $P_y = 25$ mm. The simulated S_{21} of this unit cell is shown in Figure 4.76.

From Figure 4.76 it can be seen that the infinite FSS resonates at around 2.8 GHz. The insertion loss of the FSS is very small. At this stage the substrate is not included in the simulation, but for the real-world case the FSS should be printed on a commercially available substrate. Moreover, bias circuitry should be carefully designed to switch

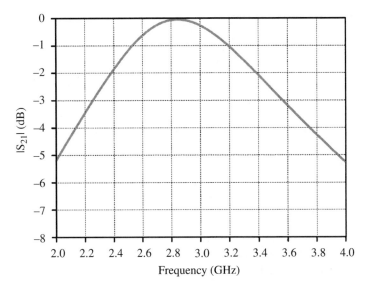

Figure 4.76 Simulated S_{21} of the passive FSS.

the states of the PIN diodes. Here we use a basing structure where FSS and PIN diodes are on the opposite sides of a very thin substrate. Specifically, a passive copper FSS is etched on one side of the substrate, while on the other side identical PIN diodes are column-by-column connected through etched strip lines. In this way, each FSS column can be individually controlled by applying DC voltages to the diodes. In view of this, the final AFSS consisting of a copper FSS layer, a thin substrate, and a bias line layer is designed and simulated in CST using the same setup as that of the previous passive FSS simulation. Some key geometrical parameters are optimised to achieve optimum transmit/reflect performances. Figure 4.77 shows the front and back views of the AFSS unit cell. As can be seen from the back view, the PIN diode is inserted in the centre of the bias line and modelled as a lumped element consisting of series/parallel RLC circuits. Careful selection of the PIN diode is crucial to the realisation of the desired AFSS performance. The most important parameters of an off-the-shelf PIN diode are the forward bias resistance R_S and the reverse bias capacitance C_j. The RF equivalent circuit of a forward biased PIN diode is a series circuit of R_s and diode package inductance L_p, while the reverse bias diode can be simplified as a series circuit of C_j and L_p (we assume the reverse parallel resistance of the PIN diode, R_p, is large enough). In this study, an Infineon-BAR6402V Silicon PIN diode is used, and its detailed electrical characteristics

Figure 4.77 Front and back view of the AFSS unit cell.

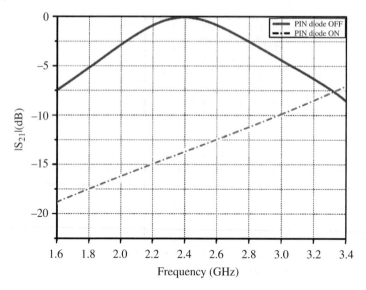

Figure 4.78 Simulated AFSS unit cell S_{21} of the ON/OFF state.

can be found in [50]. According to the parametric studies of C_j and bias line width w_f, the resonant frequency shifts from 2.4 to 1.8 GHz as the C_j varies from 0.14 to 0.6 pF. The bias line width has a minor effect on the ON state S parameters. The simulated S_{21} of the AFSS unit cell is shown in Figure 4.78.

The next step is to model a cylindrical AFSS fed by the monopole antenna. The concept of beam-switching is reconfiguring the cylindrical AFSS to mimic a corner reflector antenna. If PIN diodes in more than half of the columns are forward biased (FSS in reflect mode), while the others are reverse biased (FSS in transparent mode), the omnidirectional beam of the feed antenna can be converted to a directional beam. By alternately switching the diode ON/OFF states in each column, the beam can be steered across the entire azimuth plane. The perspective view of the CST model is shown in Figure 4.79.

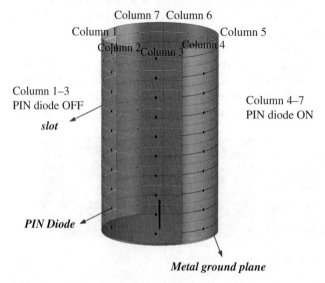

Figure 4.79 CST model of the AFSS antenna.

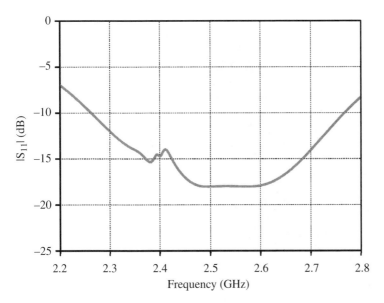

Figure 4.80 Simulated S_{11} of the AFSS antenna.

The PIN diodes in grey columns (columns 4–7) are switched ON, while those in yellow columns (columns 1–3) are switched OFF. As can be seen from the model, four columns of the AFSS cylinder act as a reflector and the other three columns are EM transparent. The relation between R_g and AFSS column number A can be denoted by $A = 2\pi R_g/P_x$, while the number of row B can be calculated by $B = H/P_y$. Here A and B are calculated to be 7 and 10, respectively. Note that R_g and H should be optimised using an equivalent metal reflector with the same R_g and H to achieve good impedance matching and radiation patterns. As shown in Figure 4.80, the simulated S_{11} of the resulting antenna has a −10 dB bandwidth from 2.26 to 2.76 GHz. From Figure 4.81, it is

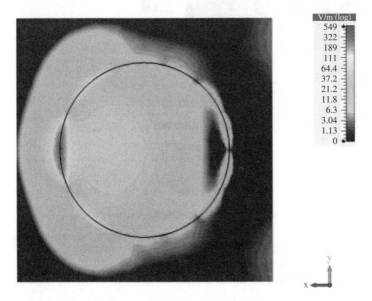

Figure 4.81 Simulated E-field distribution in the horizontal plane at 2.4 GHz.

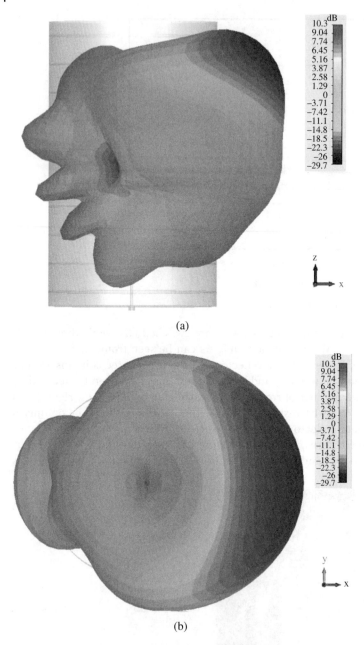

Figure 4.82 Simulated (a) side and (b) top view of the 3D radiation pattern at 2.4 GHz.

evident that most of the EM energy is directed to the x-axis direction, thus generating a directional beam.

Figure 4.82 shows the simulated side and top views of the 3D radiation pattern at 2.4 GHz. The E- and H-plane radiation patterns are shown in Figure 4.83. It can be seen that the antenna has a realised gain of 10.3 dBi at 2.4 GHz. The main beam tilt angle is

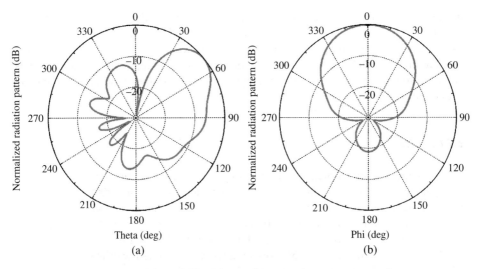

Figure 4.83 Simulated (a) E-plane and (b) H-plane radiation patterns.

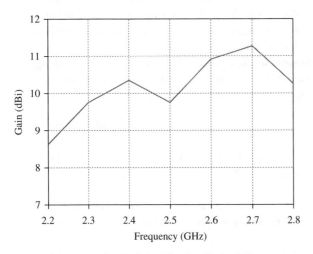

Figure 4.84 Simulated results of antenna gain vs. frequency.

39°, which resembles the beam tilt of the monopole feed. The simulated sidelobe and front-to-back ratio are −12.2 dB and 18.4 dB, respectively. Figure 4.84 illustrates that the monopole-fed AFSS antenna has a simulated gain between 8.5 and 11.2 dBi over the frequency band from 2.2 to 2.8 GHz.

4.9 Summary of the Chapter

This chapter discusses the design techniques of electronic beam-switching or beam-steerable antennas using active FSS. The operation principle is to use low-cost RF switches such as PIN diodes to reconfigure the FSS in reflection or transmissions mode, thus the beam of the antenna can be steered in the azimuth plane. More advanced techniques to realise beam-steering in both azimuth and elevation planes are also discussed in this chapter. Compared to traditional phased arrays, the beam-reconfigurable AFSS antennas can achieve a significant reduction in the complexity, weight, and cost of smart antennas because they do not require any RF phase shifters and complicated beamforming networks.

References

1 F. Yang and Y. Rahmat-Samii. Electromagnetic band gap structures in antenna engineering. Cambridge University Press, Cambridge, New York, 2009.

2 F.-R. Yang, K.-P. Ma, Y. Qian, and T. Itoh. A uniplanar compact photonic-bandgap (UC-PBG) structure and its applications for microwave circuit. *IEEE Transactions on Microwave Theory and Techniques*, 47(8):1509–1514, Aug 1999.

3 F. Yang and Y. Rahmat-Samii. Microstrip antennas integrated with electromagnetic band-gap (EBG) structures: A low mutual coupling design for array applications. *IEEE Transactions on Antennas and Propagation*, 51(10): 2936–2946, Oct 2003.

4 F. Yang and Y. Rahmat-Samii. Reflection phase characterizations of the EBG ground plane for low profile wire antenna applications. *IEEE Transactions on Antennas and Propagation*, 51(10):2691–2703, Oct 2003.

5 D.J. Kern, D.H. Werner, A. Monorchio, L. Lanuzza, and M.J. Wilhelm. The design synthesis of multiband artificial magnetic conductors using high impedance frequency selective surfaces. *IEEE Transactions on Antennas and Propagation*, 53(1):8–17, Jan 2005.

6 D. Sievenpiper, L. Zhang, R.F.J. Broas, N.G. Alexopolous, and E. Yablonovitch. High-impedance electromagnetic surfaces with a forbidden frequency band. *IEEE Transactions on Microwave Theory and Techniques*, 47(11):2059–2074, Nov 1999.

7 B.A. Munk. *Frequency Selective Surfaces: Theory and Design*. John Wiley & Sons, Inc., Hoboken, NJ, 2000.

8 T.K. Wu (ed.). *Frequency Selective Surface and Grid Array*. Wiley, New York, 1995.

9 I. Anderson. On the Theory of Self-Resonant Grids. Bell System Technical Journal, 54:1725–1731, 1975.

10 N. Amitay, V. Galindo, and C.P. Wu. *Theory and Analysis of Phased Array Antennas*. John Wiley & Sons, Chichester, 1972.

11 C.-H. Tsao and R. Mittra. A spectral-iteration approach for analyzing scattering from frequency selective surfaces. *IEEE Transactions on Antennas and Propagation*, 30:303–308, 1982.

12 P. Van Den Berg. Iterative computational techniques in scattering based upon the integrated square error criterion. *IEEE Transactions on Antennas and Propagation*, 32:10631071, 1984.

13 M. Mokhtar and E.A. Parker. Iterative computation of currents in frequency-selective surfaces of finite size. *Electronics Letters*, 3:221–222, 1989.

14 R. Mittra, C.H. Chan, and T. Cwik. Techniques for analyzing frequency selective surfaces – a review. *Proceedings of the IEEE*, 76:1593–1615, 1988.

15 M. Ohira, H. Deguchi, M. Tsuji, and H. Shigesawa. Multiband single-layer frequency selective surface designed by combination of genetic algorithm and

geometry-refinement technique. *IEEE Transactions on Antennas and Propagation*, 52(11):2925–2931, Nov 2004.

16 D.J. Kern, D.H. Werner, A. Monorchio, L. Lanuzza, and M.J. Wilhelm. The design synthesis of multiband artificial magnetic conductors using high impedance frequency selective surfaces. *IEEE Transactions on Antennas and Propagation*, 53(1):8–17, Jan 2005.

17 B.A. Munk, P. Munk, and J. Pryor. On designing Jaumann and circuit analog absorbers (CA absorbers) for oblique angle of incidence. *IEEE Transactions on Antennas and Propagation*, 55(1):186–193, Jan 2007.

18 J.A. Bossard et al. Tunable frequency selective surfaces and negative-zero-positive index metamaterials based on liquid crystals. *IEEE Transactions on Antennas and Propagation*, 56(5):1308–1320, May 2008.

19 E. Parker. The gentleman's guide to frequency selective surfaces. In *17th QMW Antenna Symposium*, pages 1–18, Kent Academic Repository, 1991.

20 M.N. Jazi and T.A. Denidni. Frequency selective surfaces and their applications for nimble-radiation pattern antennas. *IEEE Transactions on Antennas and Propagation*, 58:2227–2237, 2010.

21 A. Edalati and T.A. Denidni. Frequency selective surfaces for beam-switching applications. *IEEE Transactions on Antennas and Propagation*, 61(1):195–200, Jan 2013.

22 M. Niroo-Jazi and T.A. Denidni. Electronically sweeping-beam antenna using a new cylindrical frequency-selective surface. *IEEE Transactions on Antennas and Propagation*, 61(2):666–676, Feb 2013.

23 B. Sanz-Izquierdo, E. A. Parker, J.-B. Robertson, and J. C. Batchelor. Tuning technique for active FSS arrays. *Electronics Letters*, 45(22), p. 1107, 2009.

24 B. Sanz-Izquierdo, E.A. Parker, and J.C. Batchelor. Dual-band tunable screen using complementary split ring resonators. *IEEE Transactions on Antennas and Propagation*, 58(11):3761–3765, Nov 2010.

25 B. Sanz-Izquierdo, E.A. Parker, and J.C. Batchelor. Switchable frequency selective slot arrays. *IEEE Transactions on Antennas and Propagation*, 59(7):2728–2731, Jul 2011.

26 B. Liang, B. Sanz-Izquierdo, E.A. Parker, and J.C. Batchelor. Cylindrical slot FSS configuration for beam-switching applications. *IEEE Transactions on Antennas and Propagation*, 63(1):166–173, Jan 2015.

27 C. Gu, B. Sanz-Izquierdo, S. Gao, J.C. Batchelor, E.A. Parker, F. Qin, G. Wei, J. Li, and J. Xu. Dual-band electronically beam-switched antenna using slot active frequency selective surface. *IEEE Transactions on Antennas and Propagation*, 65(3):1393–1398, Mar 2017.

28 C. Gu, S. Gao, B. Sanz-Izquierdo, E.A. Parker, F. Qin, H. Xu, J.C. Batchelor, X. Yang, and Z. Cheng. 3D-coverage beam-scanning antenna using feed array and active frequency selective surface. *IEEE Transactions on Antennas and Propagation*, 65(11): 5862–5870, 2017.

29 Z. Zaharis, E. Vafiadis, and J.N. Sahalos. On the design of a dual-band base station wire antenna. *IEEE Antennas and Propagation Magazine*, 42:144–151, 2000.

30 R.L. Li, T. Wu, B. Pan, K. Lim, J. Laskar, and M.M. Tentzeris. Equivalent-circuit analysis of a broadband printed dipole with adjusted integrated balun and an array for base station applications. *IEEE Transactions on Antennas and Propagation*, 57(7):2180–2184, Jul 2009.

31 R.C. Hansen. *Phased Array Antennas*. John Wiley & Sons Inc., Hoboken, NJ, 2009.

32 C.-H. Ko, K.M. J. Ho, and G.M. Rebeiz. An electronically-scanned 1.8–2.1 GHz base-station antenna using packaged high-reliability RF MEMS phase shifters. *IEEE Transactions on Microwave Theory and Techniques*, 61(2):979–985, Feb 2013.

33 F. Falcone et al. Babinet principle applied to the design of metasurfaces and meta-materials. *Physical Review Letters*, 93(19), Nov 2004.

34 F. Tefiku and C.A. Grimes. Design of broad-band and dual-band antennas comprised of series-fed printed-strip dipole pairs. *IEEE Transactions on Antennas and Propagation*, 48(6):895–900, Jun 2000.

35 K. Ogawa and T. Uwano. A variable tilted fan beam antenna for indoor base stations. In *Antennas and Propagation Society International Symposium*, pages 332–335, AP-S. Digest, 1994.

36 S.-W. Lu, T.-F. Huang, and P. Hsu. CPW-fed slot-loop coupled patch antenna on narrow substrate. *Electronics Letters*, 35:682–683, 1999.

37 F. Bayatpur and K. Sarabandi. A tunable metamaterial frequency-selective surface with variable modes of operation. *IEEE Transactions on Microwave Theory and Techniques*, 57:1433–1438, Jun 2009.

38 M. Safari, C. Shafai, and L. Shafai. X-band tunable frequency selective surface using MEMS capacitive loads. *IEEE Transactions on Antennas and Propagation*, 63:1014–1021, 2015.

39 D. Cure, T.M. Weller, and F.A. Miranda. Study of a low-profile 2.4-GHz planar dipole antenna using a high-impedance surface with 1D varactor tuning. *IEEE Transactions on Antennas and Propagation*, 61:506–515, 2013.

40 W. Hu, R. Dickie, R. Cahill, H. Gamble, Y. Ismail, V. Fusco, D. Linton, N. Grant, and S. Rea. Liquid crystal tunable mm wave frequency selective surface. *IEEE Microwave and Wireless Components Letters*, 17:667–669, 2007.

41 C. Gu, S. Gao, B. Sanz-Izquierdo, E.A. Parker, W. Li, X. Yang, and Z. Cheng. Frequency-agile beam-switchable antenna. *IEEE Transactions on Antennas and Propagation*, 65(8):3819–3826, Aug 2017.

42 C. Mias. Varactor-tunable frequency selective surface with resistive-lumped-element biasing grids. *IEEE Microwave and Wireless Components Letters*, 15(9):570–572, Sep 2005.

43 F. Costa, A. Monorchio, S. Talarico, and F.M. Valeri. An active high-impedance surface for low-profile tunable and steerable antennas. *IEEE Antennas and Wireless Propagation Letters*, 7:676–680, 2008.

44 P. Baccarelli, C. Di Nallo, F. Frezza, A. Galli, and P. Lampariello. The role of complex waves of proper type in radiative effects of nonreciprocal structures. In *Proceedings of the 1997 IEEE MTT-S*, volume 2, pages 491–494, IEEE, 1997.

45 A. Lai, K. Leong, and T. Itoh. Leaky-wave steering in a two-dimensional metamaterial structure using wave interaction excitation. *In Proceedings of the International Microwave Symposium, IEEE*, pages 1643–1646, 2006.

46 W. Pan, C. Huang, P. Chen, M. Pu, X. Ma, and X. Luo. A beam steering horn antenna using active frequency selective surface. *IEEE Transactions on Antennas and Propagation*, 61(12):6218–6223, Dec 2013.

47 L. Boccia, I. Russo, G. Amendola, and G. Di Massa. Multilayer antenna-filter antenna for beam-steering transmit-array applications. *IEEE Transactions on Microwave Theory and Techniques*, 60(7):2287–2300, Jul 2012.

48 M. Sazegar et al. Beam steering transmitarray using tunable frequency selective surface with integrated ferroelectric varactors. *IEEE Transactions on Antennas and Propagation*, 60(12):5690–5699, Dec 2012.

49 R.J. Mailloux. *Phased Array Antenna Handbook*. Artech House, Boston, 2005.

50 Infineon Technologies. *Silicon PIN Diode*, BAR64-02V datasheet, June 2013. [Online]. Available: https://www.infineon.com/cms/en/product/rf-wireless-control/rf-diode/rf-pin-diode/antenna-switch/bar64-02v/.

5

Beam Reconfigurable Reflectarrays and Transmitarrays

5.1 Introduction

Reflectarrays and transmitarrays have received a significant amount of research interest recently. Reflectarrays combine the advantages of reflector antennas and phased arrays. A conventional microstrip reflectarray consists of a printed array (e.g. of patch antennas) illuminated by at least one feed antenna such as a circular horn. Each radiating element is designed to provide a pre-adjusted phase to form a focused beam to the desired direction. At its operation frequencies, a reflectarray antenna can substitute the traditional parabolic reflector with a lightweight planar structure. A transmitarray antenna also consists of feed antenna(s) and an array of unit cells. Different from the reflectarrays, the array of unit cells transmits the waves instead of reflecting them back. Thus, a transmitarray is referred to as a 'planar lens' in some literature. The transmitarrays place the feed antenna directly in front of the aperture without incurring any blockage losses. This is especially advantageous when there is a need to use multiple feeds to obtain multi-beams or beam-switching.

For a low-cost smart antenna design, reflectarrays and transmitarrays are quite attractive for two reasons. First, they employ spatial feeding instead of using traditional feed networks. This is an important advantage as it enables a significant reduction in system complexity and cost, particularly when there is a need to use a large number of antenna elements to achieve high directivity. One well-known problem of traditional microstrip array antenna is the high loss in the feed network at high frequencies, such as mm-wave frequencies, which leads to significant degradation of the efficiency of the array antenna. Thanks to space feeding, reflectarrays and transmitarrays eliminate the losses of feed networks. Second, reflectarrays and transmitarrays are easy to manufacture using low-cost printed circuit board (PCB) technologies and can be designed to have planar structures as well as conforming to the bodies of satellites, air planes or vehicles.

The use of transmitarrays to develop cost-effective 5G massive MIMO has already been reported and patented [1–3]. To obtain beam-steering, the traditional approach is to incorporate phase shifters on the antenna element. However, phase shifters are expensive, especially at high frequencies (e.g. mm-wave frequencies). Thus, the techniques presented in this chapter focus on using low-cost components, including PIN and varactor diodes, to design a beam reconfigurable reflectarray or transmitarray. A brief introduction to the basic principles of reflectarrays and transmittarrays is given first,

Low-cost Smart Antennas, First Edition. Qi Luo, Steven (Shichang) Gao, Wei Liu, and Chao Gu.
© 2019 John Wiley & Sons Ltd. Published 2019 by John Wiley & Sons Ltd.

followed by discussions of design techniques of beam-reconfigurable reflectarrays and transmitarrays. Many recent examples of beam-reconfigurable reflectarrays and transmitarrays are given and discussed. One case study is provided to explain the techniques and step-by-step procedures to design a multi-beam transmitarray.

5.2 Reflectarray and Transmitarray Design Fundamentals

This section briefly introduces the design fundamentals for reflectarrays and transmitarrays. The aim of this section is to outline the operation principles of these two types of space-fed array antenna and related design basics. The design of reflectarrays and transmitarrays is a comprehensive topic. There are some books that focus on the analysis and design of reflectarray [4, 5] and transmitarray [6] antennas, and interested readers can refer to these three books for more details.

5.2.1 Reflectarrays

A printed microstrip reflectarray consists of a printed array antenna that is illuminated by the feed antenna. Each radiating element is designed to provide a pre-adjusted phase to form a focused beam along the desired direction. At its operation frequencies, a reflectarray shows comparable radiation performances compared to a traditional parabolic reflector with a lightweight planar structure. Reflectarrays with wide beam-scanning angle range [7], ultra-wide bandwidth [8], and integrated feed [9] have been recently reported in the literature.

Figure 5.1 shows the configuration of a central-fed printed microstrip reflectarray. The radiating elements of the reflectarray receive incident waves from the feed antenna through space feeding and re-radiate the incident field with pre-designed phases. The operating mechanism is similar to traditional reflector antennas except that it is realised by using the concept of phased array antennas.

To reduce the profile of the conventional reflectarray, the folded reflectarray has been proposed to provide a low profile solution [10]. Figure 5.2 shows the configuration of a folded reflectarray. The folded reflectarray usually consists of a feed antenna, an array of radiating elements, and a polariser grid. As can be seen from Figure 5.2, for the folded reflectarray the feed horn is linearly polarised and the radiated wave from the feed horn is reflected back by the polariser. Then, the reflected wave is incident on the reflectarray and re-radiates with the electric field rotated by 90°. This is equivalent to having an imaginary feed above the centre of the reflectarray.

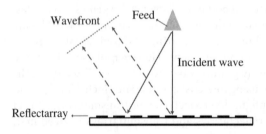

Figure 5.1 The configuration of a printed microstrip reflectarray.

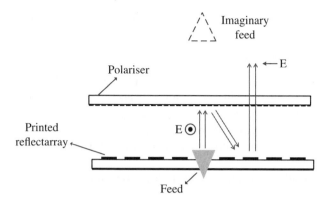

Figure 5.2 The configuration of a folded reflectarray.

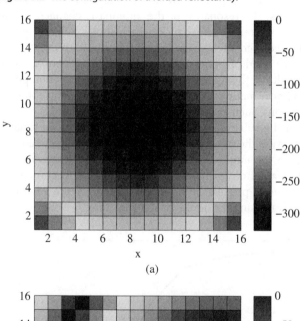

(a)

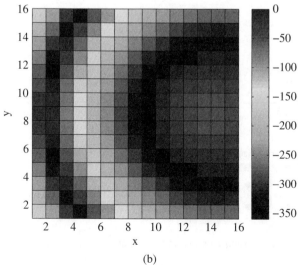

(b)

Figure 5.3 Examples of the required phase distribution for a planar reflectarray with focused beam at (a) broadside and (b) 20° off the broadside.

The required phase for each radiating element is calculated by using the formula given in [4]

$$\phi_R = k_0(d_i - (x_i \cos \varphi_b + y_i \sin \varphi_b) \sin \theta_b) \tag{5.1}$$

where ϕ_R is the phase of the reflection coefficient of the antenna element i, k_0 is the phase constant in vacuum, (x_i, y_i) are the coordinates of the array element i, d_i is the distance from the phase centre of the feed to the radiating element, and (θ_b, φ_b) is the targeted scan angles of the beam in the spherical coordinate system. Figure 5.3 shows some examples of the required phase distribution for a planar reflectarray with beam at different angles.

There are several approaches to adjust the phase of the reflectarray unit cells and they are summarised below:

1. Modify the size of the radiating elements [11, 12]. By varying the size of the radiating element, e.g. microstrip patches, the resonance of the antenna is changed so the radiating element has different scattering impedance, which shows different phase delays when re-radiating the received RF signals. Figure 5.4 shows an example of a microstrip square patch of varied size and its corresponding reflection phase delays. As shown, when the width of the square patch increases, the reflection phase increases as well but not in a linear manner. A more linear response can be obtained by using multilayer configurations, such as stacked patch antennas [13].
2. Change the characteristics of the substrate [14]. Recent developments on tunable substrates allows the permittivity of the substrate to be varied. For example, by adjusting the bias voltage applied to the liquid crystals, its electrical properties can be changed. Thus, the microstrip antennas printed on such a substrate become electrically reconfigurable, and different phase delays can be obtained. Figure 5.5 shows the configuration of a reflectarray using the liquid crystal substrate.

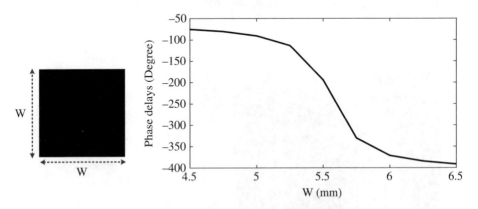

Figure 5.4 Microstrip square patch of varied size and its corresponding reflection phase delays.

Figure 5.5 The configuration of a reflectarray using a liquid crystal substrate.

3. Load the radiating element with open-circuited microstrip stubs of varied length. This method uses identical antenna elements with variable length phase delay lines attached. The phase delay line is an open microstrip line which reflects the received RF signal back to the radiating element, and its electrical length is proportional to the phase delays. Figure 5.6 shows an example of a microstrip square patch with delay lines of varied length and its corresponding reflection phases. As can be seen, a linear phase response can be achieved by using this approach. To allow more space to place the phase delay line and introduce additional circuitry to obtain a reconfigurable design, the aperture coupled feed method is a preferred configuration. Figure 5.7 shows the configuration of the aperture coupled patch. As shown, the patch and the feed line are electromagnetically coupled through a slot etched on the ground plane. Another advantage of this type of configuration is that the ground plane can shield the unwanted radiation from the microstrip lines and the other RF components, which makes the design more robust.
4. Rotate the radiating element. This approach rotates the identical circularly polarised radiating elements to obtain the phase delay [15]. By rotating the radiating element by an angle of ψ, the reflected wave is delayed by 2ψ. This method is suitable for the design of circularly polarised reflectarrays.

5.2.2 Transmitarrays

The configuration of the transmitarray antenna is very similar to that of reflectarrays. It consists of an illuminating feed source and a transmitting surface, which is typically

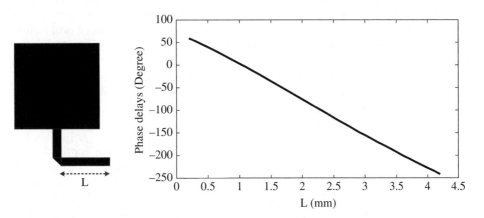

Figure 5.6 Microstrip square patch with delay lines of varied length and corresponding reflection phase delays.

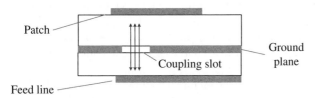

Figure 5.7 The configuration of the aperture coupled patch.

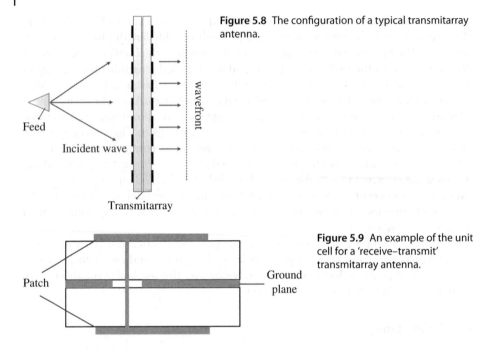

Figure 5.8 The configuration of a typical transmitarray antenna.

Figure 5.9 An example of the unit cell for a 'receive–transmit' transmitarray antenna.

a planar array. Different from reflectarrays, the transmitarray transforms the incident wave into outgoing waves instead of reflecting them back. In some literature, the transmitarray is also mentioned as a planar lens antenna [16]. Compared to the traditional dielectric lens, the transmitarray has a planar configuration with a reduced profile, low fabrication cost, and light weight, especially at microwave frequencies [17]. Figure 5.8 shows the configuration of a typical transmitarray antenna.

Compared to the reflectarray, the transmitarray has a higher profile because it uses additional layers. However, as the RF signals are not reflected back, the transmitarray has no blockage effect. Meanwhile, it is less sensitive to surface errors [4]. Generally speaking, there are two approaches to designing a transmitarray.

1. The first approach is based on the array antenna [18, 19]. Using this approach, one array antenna receives the RF signal from the feeding source and transmits it to another array antenna which re-transmits the RF signal with pre-adjusted phase to form a cophasal beam in the far-field. Figure 5.9 shows an example of the unit cell for such a 'receive–transmit' configuration. In this example, two microstrip patches are interconnected using a signal via. Alternatively, an aperture-coupled configuration can also be used in which the feed lines of the two patches are electromagnetically coupled through a slot etched on the shared ground plane [4]. Figure 5.10 shows the configuration of the transmitarray unit cell that has two patch apertures coupled to each other. The feed lines can be either printed on the same layer of the patch or be placed at different layers, depends on the available space for accommodating the microstrip lines.
2. The second approach is based on the FSS or metamaterial (MTM) [1, 20, 21]. This approach uses FSSs as free-space phase shifters to obtain the desired phase delays to perform the beamforming. Because of the low insertion loss and wide bandwidth

Figure 5.10 Exploded view of the transmitarray unit cell using the aperture coupling technique.

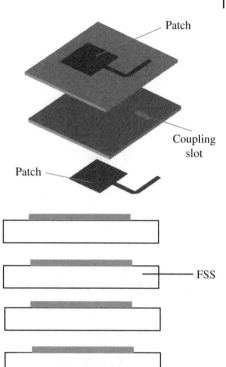

Patch

Coupling slot

Patch

Figure 5.11 An example of the unit cell for an FSS-based transmitarray antenna.

FSS

of the FSSs, the resulted transmitarray has a wide bandwidth and high efficiency. However, to achieve a phase shift range of 360°, multiple FSS layers spaced by quarter-wavelengths are needed, which leads to an increased profile. Figure 5.11 presents the configuration of a FSS-based transmitarray antenna. Note that each FSS layer does not necessarily need to be the same.

5.3 Beam Reconfigurable Reflectarrays

5.3.1 Multi-feed Reflectarray

Multi-beam reflectarrys can be realised by using multiple feeds and designing the reflectarray as a bifocal antenna. By incorporating an RF feed switching unit, the multi-beam reflectarray is extended to a beam-switching reflectarray. This approach represents a classic method to realise beam-switching with low system complexity and overall cost.

A multi-beam folded reflectarray was reported in [22] and the concept of this type of multi-beam reflectarry is presented in Figure 5.12. By switching between the feed antennas, the reflectarray reconfigures its beam to different angles. As shown in [22], the beam of the folded reflectarray can be switched between ±13.5° with beamwidth between 3° and 3.5° and sidelobe level better than −18 dB.

To design a bifocal reflectarray, using the ray racing theory, the following condition needs to be satisfied [23]

$$\Delta r \times \sin \theta_{out} + \phi_1 = \Delta r \times \sin \theta_{in} + \phi_2 \tag{5.2}$$

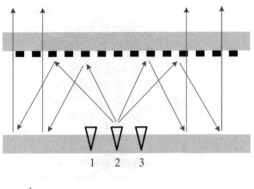

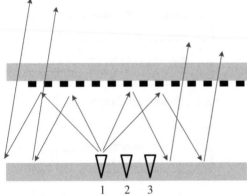

Figure 5.12 Multi-beam operation of the multi-feed folded reflectarray.

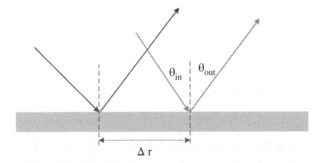

Figure 5.13 Multi-beam operation of the multi-feed folded reflectarray.

where Δr is a small distance between two parallel rays, θ_{in} is the incident angle of the ray, and θ_{out} is the reflected angle of the ray. ϕ_1 and ϕ_2 represent the phase delay of the reflectarray for ray 1 and ray 2, respectively. These parameters are shown in Figure 5.13. When Δr is close to zero

$$\frac{\partial \varphi}{\partial r} = \sin \theta_{out} - \sin \theta_{in} \tag{5.3}$$

Figure 5.14 The configuration of the multi-feed reflectarray reported in [24].

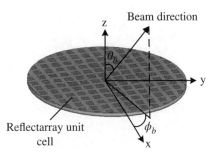

Equation 5.3 only considers the one-dimensional ray reflection. For a two-dimensional case, Equation 5.3 is re-written as

$$\frac{\partial \varphi}{\partial r} = \sin \theta_{out} \times \cos \phi_{out} - \sin \theta_{in} \times \cos \phi_{in} \tag{5.4}$$

$$\frac{\partial \varphi}{\partial r} = \sin \theta_{out} \times \sin \phi_{out} - \sin \theta_{in} \times \sin \phi_{in} \tag{5.5}$$

By performing the phase-only synthesis, a multi-feed reflectarray with independent beams and improved radiation pattern can be obtained. For example, a multi-feed printed reflectarray with three feed horns and three simultaneous shaped beams was presented in [24]. The configuration of this reflectarray is shown in Figure 5.14. After performing the phase-only synthesis, the beam shape is improved by applying an optimisation technique that is based on the intersection approach method [25].

The phase-only synthesis and the intersection approach method are based on the fact that the radiation pattern of the reflectarray can be expressed by using two-dimensional fast Fourier transform (2D-FFT). It shown in [24] that the radiation integrals of the co-polarisation throughout the reflectarray aperture can be expressed as

$$P_x^X \approx N_x \times N_y \times K \times [IDFT2(\rho_{xx}(m, n)) * IDFT2(E_{xinc}^X(m, n))] \tag{5.6}$$

where P_x^X is the radiation integrals of the co-polarisation assuming that the feed is linearly polarised in the x direction, IDFT2 represents the two-dimensional inverse discrete Fourier transform, ρ_{xx} is the reflection coefficient, and $E_{xinc}^X(m, n)$ is the copolar incident field on the reflector surface. K is a constant and is defined as

$$K = \Delta x \times \Delta y \times sinc(\frac{k_0 u \Delta x}{2}) \times sinc(\frac{k_0 v \Delta y}{2}) \tag{5.7}$$

where k_0 is the propagation constant, and

$$u = \sin \theta \times \cos \phi \tag{5.8}$$

$$v = \sin \theta \times \sin \phi \tag{5.9}$$

To design the reflectarray, stacked patches were used as the unit cell. The configuration of the unit cell is shown in Figure 5.15. The advantages of using stacked patches include compact size and wide bandwidth. Arlon CuClad 233LX ($\varepsilon_r = 2.33$, $\tan \sigma = 0.0013$) with

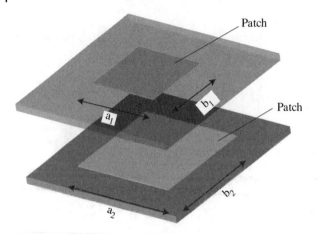

Figure 5.15 The configuration of the reflectarray unit cell used in [24].

thickness 0.787 mm is chosen as the substrate, and the layers were bonded by CuClad 6250 bonding film ($\varepsilon_r = 2.32$, $\tan \sigma = 0.0013$) with thickness 0.037mm. The phase delay of the unit cell is achieved by varying the size of the patch. Using this unit cell, more than 500° phase shift range can be realised. For each unit cell, the sizes of the two stacked patches are defined to have the same ratio

$$a_2 = 0.75a_1, b_2 = 0.75b_1 \tag{5.10}$$

5.3.2 Reflectarray with RF Switches

Using a single feed, a typical reconfigurable reflectarray antenna consists of an array of reflecting elements with electronically controllable phase shifters or an array of reconfigurable reflecting elements. The use of the digital-phase shifters has the advantage of providing higher phase quantisation resolutions and being compatible with many commercially available digital control circuits. However, the cost of the phase shifter is high, especially at high frequencies, which makes this approach not suitable for low-cost smart antenna design. One approach to realise a low-cost design is to introduce RF switches, such as PIN diodes and MEMS, to the phase delay lines of the radiating element. This is equivalent to having a low-bit phase shifter. By doing this, the unit cell of the reflectarray can have a simple structure and less complicated control circuit. This is a trade-off between having a low-cost solution and a high-performance array: due to the phase quantisation error, there is an increased sidelobe level and directivity decrease [26].

Figure 5.16 shows an example of introducing one RF switch to the phase delay line of the reflectarray element. The radiating element is a microstrip patch and the required phase delay is obtained by using a microstrip feed line of varied length. Through controlling the states of the RF switches, the electrical length of the phase delay line can be changed, which is equivalent to a 1-bit phase shifter. One of the most common and cost-effective RF switches is the RF PIN diode. RF PIN diodes exhibit fast switching time with reasonable insertion loss and isolation. Currently, RF PIN diodes for mm-wave applications are commercially available, and PIN diodes have been successfully implemented in the design of mm-wave reflectarrays.

Figure 5.16 Introducing an RF switch to the phase delay line of the reflectarray element.

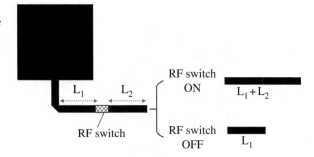

An electronically beam-steerable reflectarray antenna operating in the 60 GHz band was presented in [27]. This reflectarray consists of 160×160 microstrip square patches. A single-bit digital phase shifter using PIN diodes was used to electronically reconfigure the phase of each radiating element. Since the phase only has two states, after calculating the desired phase delays ϕ_R (using Equation 5.1) for each element, they are quantised as

$$\phi_q = +\frac{\pi}{2}, \quad \text{when} \quad 0 \le \phi_R (modulo\ 2\pi) < \pi \tag{5.11}$$

$$\phi_q = -\frac{\pi}{2}, \quad \text{when} \quad \pi \le \phi_R (modulo\ 2\pi) < 2\pi \tag{5.12}$$

where ϕ_q is the phase after being quantised and assigned to the PIN states. The main beams of this reflectarray can be scanned in both the azimuth and elevation planes within $\pm 25°$ with 3-dB beamwidth from 0.55° to 0.63° with low sidelobe level. To control the states of each PIN diode, active matrix architecture was used, which is shown in Figure 5.17. The circuitry is placed at the back side of the reflectarray. The PIN diode is driven by the outputs of the latches. To reduce the complexity of the control circuit and minimise the number of control lines, which is especially critical for an array antenna of large element number, the parallel control data is sent row by row. A control data distributor uses an FPGA to send parallel control data to the latches and the shift register

Figure 5.17 The control circuitry of the 1-bit electronically beam-steerable reflectarray presented in [27].

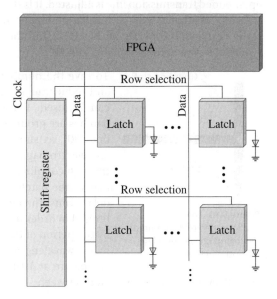

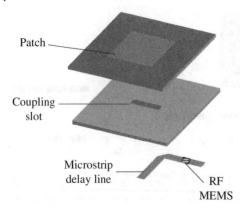

Patch

Coupling slot

Microstrip delay line

RF MEMS

Figure 5.18 Configuration of the aperture-coupled microstrip patch antenna element with MEMS switches integrated on the phase delay line presented in [29].

lets the latches retain the control data. As reported in [27], using this configuration, with a clock at 16.7 MHz, it took about 10 μs to send all the control data. It is also estimated that considering half of the PIN diodes are in the OFF state (a total of 25,600 PIN diodes were used), the power dissipated by the PIN diodes is about 234 W.

A reconfigurable reflectarray with single-bit phase resolution for Ku-band satellite applications was reported in [28]. In this design, four PIN diodes were symmetrically embedded in the radiating element and by switching the states of the integrated PIN diodes in each unit cell the beam of the reflectarray was electronically scanned with an angular range of ±40°. Besides PIN diodes, RF MEMS switches can also be used to control the phase of the reflectarray element. RF MEMS switches have the advantage of low insertion loss, high isolation, low DC power consumption, and linear response within a wide frequency range. Figure 5.18 shows the configuration of the aperture-coupled microstrip patch antenna element used in the beam-switching reflectarray reported in [29]. The MEMS switches are monolithically integrated on the open-ended microstrip phase delay lines and by changing the states of the MEMS, the equivalent length of the open-ended transmission line is adjusted. It is shown that the insertion loss of the MEMS switch is less than 0.5 dB with isolation higher than 10 dB up to 30 GHz. This reflectarray is designed to operate at 26.5 GHz and the main beam of the reflectarray can be switched between the broadside and 40°.

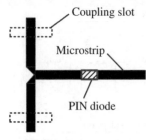

Coupling slot

Microstrip

PIN diode

Figure 5.19 The configuration of the common microstrip phase delay line proposed in [30].

To save the number of RF switches and control devices, and reduce the manufacturing cost, the subarray concept can be used. For example, the feed lines of a pair of patches are grouped to a common microstrip phase delay line [30], as shown in Figure 5.19. The microstrip lines are electromagnetically coupled to rectangular patches through coupling slots. With a shared phase delay line, the pair of the patches is configured to have the same phase delay. Thus, the beam-steering angle of the reflectarray is limited in order to avoid the grating lobes, which is caused by the period of the subarray becoming much greater than half a wavelength. In [30], the reflectarray was designed to operate at 10.4 GHz (X-band), and the main beam of the reflectarray can be switched between −5°, 0°, and +5°. To alleviate the effect of the grating lobes and improve the beam-scanning performance of the reflectarray, an irregular lattice can be used [31].

Reconfigurable design can also be achieved by introducing RF switches directly on the radiating element, such as integrating MEMS to the ring-type element [32]. However, for better system integration, it is recommended that the RF switches and radiating element are placed on the different layers, e.g. using the configuration shown in Figure 5.18. Although such a configuration leads to multilayer structures and increases the fabrication complexity, it does not significantly increase the fabrication cost. The advantages include the following:

- Improvement of the radiation performance of the antenna by avoiding unwanted radiation. Placing the phase delay lines and related biasing circuitry on a different layer to the radiating element, the spurious radiation from the microstrip and bias lines can be shielded by the ground plane of the antenna.
- Allowing more spaces to accommodate the RF switches and circuit, especially for mm-wave applications. It is known that to avoid the grating lobes, the spacing between adjacent elements should be close to a half wavelength at the frequency of interest. Thus, it is not feasible to place a large number of RF switches and print the bias lines without affecting the radiation performance of the antenna. In this case, it is highly desirable to place the biasing circuitry on a different layer.

5.3.3 Reflectarray with Tunable Components

As presented in the previous section, the use of RF switches is equivalent to the use of a low-bit phase shifter, which can only provide two states of phase delays if only one RF switch is integrated on the unit cell. This decreases the directivity and limits the beam-scanning capability of the reflectarray due to quantisation errors. To reach a continuous phase tuning, the varactor can be used.

The varactor has voltage-dependent capacitance. This characteristic can be used to load the microstrip delay lines of a reflectarray. Figure 5.20a shows an open-ended microstrip transmission line loaded by a varactor. This microstrip is printed on a Roger 4003C substrate ($\varepsilon_r = 3.55$, $\tan \sigma = 0.0027$) with thickness 0.3mm. Figure 5.20b is the simulated phase of the reflection coefficient at 10 GHz when the capacitance of the varactor varies. As shown, when the capacitance changes from 0.1 to 0.5 pF, the phase shift of the reflected signal is about 80°. A larger phase shift range can be obtained by further increasing the capacitance of the varactor through changing the applied voltage on the varactor.

The results shown in Figure 5.20 are for an ideal case when the varactor is only considered as a capacitor. In practical design, it must take into account the package parasitic effects and the diode losses. Figure 5.21 shows the equivalent circuit model for a varactor [33]. In this figure, $C_j(V)$ is the variable junction capacitance of the diode die, $R_s(V)$ is the variable series resistance of the diode die, C_p is the fixed parasitic capacitance, and L_p is the parasitic inductance.

Figure 5.22 shows the configuration of the unit cell of a single varactor based reconfigurable reflectarray [34]. The unit cell is an aperture-coupled patch antenna operating at the X-band and the varactor is used to load the microstrip delay line. The varactor used is a Microsemi MV31011-89 varactor diode, whose capacitance can be adjusted from

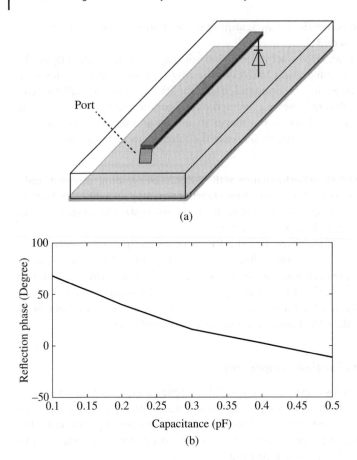

(a)

Figure 5.20 (a) The microstrip line loaded by a varactor. (b) Simulated reflection phase of the microstrip line at 10 GHz when the capacitance of the varactor varies.

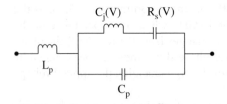

Figure 5.21 The equivalent circuit model of the varactor.

0.2 to 2 pF when the biasing voltage is adjusted from 20 to 0 V. At the resonance of the patch, a phase tuning range of 320° is obtained.

A reflectarray prototype of 3×15 elements was designed and fabricated using the unit cell shown in Figure 5.22. An array of 12 AD5764R DACs, each of which contains four high-accuracy 16-bit DACs, is used to produce the voltages to control the varactor diodes and a microcontroller is used to serially send the desired voltage levels to the DACs. Figure 5.23 shows a photo of the varactor-loaded microstrip phase delay line with the biasing lines and DAC board. By electronically controlling the biasing voltages

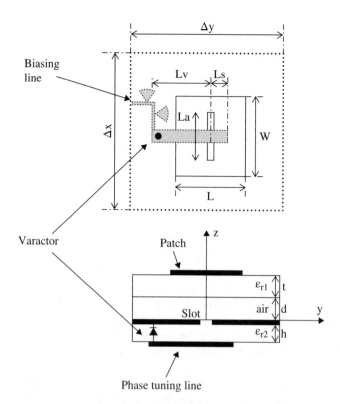

Figure 5.22 The configuration of the varactor loaded reflectarray element [34]. Reproduced with permission of ©EurAAP.

Figure 5.23 The configuration of the varactor loaded reflectarray element [34]. Reproduced with permission of ©EurAAP.

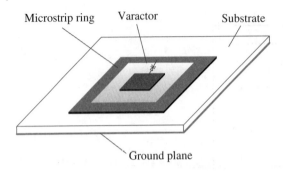

Microstrip ring   Varactor   Substrate

Ground plane

of the varactor, the main beam of the reflectarray can be switched from $-20°$ to $+20°$ in steps of $10°$.

A dual-band electronically beam-shaping reflectarray using a ring-type patch loaded by a single varactor diode that connects the gap between the ring and the patch is presented in [35]. Figure 5.24 shows the configuration of this unit cell. The unit cell is a dual-band element, with one resonance determined by the square patch and another resonance determined by the ring. The mutual coupling between the ring and the patch can be used to tune the operation frequencies of this unit cell. Thus, a varactor is placed across the gap between the ring and the square patch. By varying the capacitance of the varactor, the resonance of the unit cell is changed, which leads to a different phase delay response for its reflected wave.

The varactor diode used in [35] has a dynamic range of 0.1 to 0.6 pF, and more than $330°$ phase shift can be achieved at both frequency bands. It is noted that the polarity of the diode needs to be aligned with the electric field of the incident wave, so the reflectarray is single linearly polarised. To obtain a dual polarised design, instead of only using one varactor, multiple varactors can be symmetrically placed around the patch [36]. The operation principle of the unit cell shown in Figure 5.24 can be explained as introducing one variable capacitor to a shunt LC circuit, as shown in Figure 5.25. Figure 5.25 is a simplified circuit model which assumes that the two resonances are relatively independent. Although this assumption is not accurate, this provides a good start point to develop the equivalent circuit model of this unit cell.

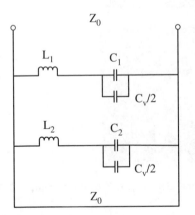

Z_0

L_1 C_1

$C_v/2$

L_2 C_2

$C_v/2$

Z_0

Figure 5.25 The simplified equivalent circuit model of the unit cell [35].

A 10×10 element reflectarray antenna was fabricated and measured in [35]. Since the varactors are mounted on the same layer of the radiating element, the biasing network for the varactor was fabricated on a separate substrate and glued to the ground plane of the reflectarray. Vias are introduced to make the connection between the biasing circuit and the varactor; thus, it is important to consider the effect of the vias by performing some necessary optimisation in the EM simulators. The measurement results show that the main beam can be steered up to $\pm 60°$ in both frequency bands with low cross-polarisations.

5.4 Beam Reconfigurable Transmitarray

The operation principle of the transmitarray is similar to the reflectarray, but its receive–transmit configuration constraints how a reconfigurable transmitarray for smart antenna applications should be designed. Compared to reflectarrays, transmitarrays have the advantage of avoiding the blockage from the feed antennas. Thus, they provide a very efficient solution to using multiple feeds to design a multi-beam or beam-switching transmitarray. Figure 5.26 shows the configuration of a multi-feed transmitarray. The transmitarray operates as a planar lens antenna and by placing multiple feed antennas on the focal arc the angle of the main beam can be switched to different angles. In the case that the focal distance is reasonably large, the feeds can be placed in parallel (as shown in Figure 5.26) instead of on the focal arc. By incorporating a feed-switching network, a low-cost beam-switching transmitarray is obtained.

Figure 5.27 shows the radiation patterns of a transmitarray operating at 13.5 GHz. The dimension of the transmitarray is 90×90 mm and it has 225 unit cells positioned in a square aperture. A linearly polarised corrugated circular horn antenna is used as the feed antenna and is placed at a distance of 90 mm above the centre of the planar array, corresponding to a focal/diameter (f/D) ratio of 1. Four feeds separated by a distance of 12.5 mm were used as the feed array and by activating the feed sequentially, the beam was switched to 15° with low sidelobes.

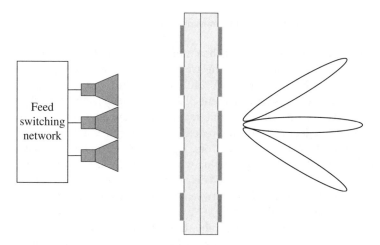

Figure 5.26 The configuration of a multi-feed transmitarray.

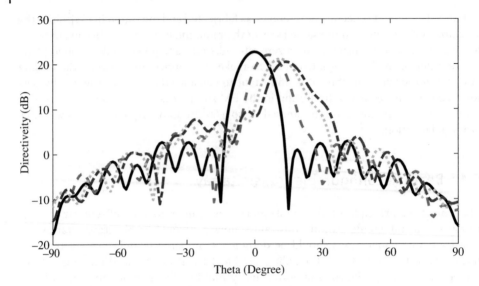

Figure 5.27 Beam-switching performance of the multi-feed transmitarray.

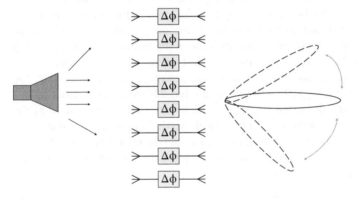

Figure 5.28 The operating principle of a beam-steerable transmitarray antenna.

A metamaterial-based transmitarray for spatial beamforming and multibeam massive MIMO was reported in [1]. The unit cell of the transmitarray is a double-layered Jerusalem cross cell separated by an air layer. A 240° phased shift was achieved by varying the length of the end cap. A seven-element SIW-fed stacked-patch is used to feed the transmitarray and the beam of the transmitarray can be scanned within −27° to +27°. Instead of using multiple feeds, a low-cost reconfigurable transmitarray can also be designed by introducing RF switches or varactors. Figure 5.28 shows the operating principle of a beam-steerable transmitarray antenna with a single feed. Different from the reflectarray, whose biasing circuitry and phase delay lines can be placed behind the ground plane, the phase delay units is placed between the receiving and transmitting panels, which leads to more design challenges.

One approach to designing a beam-reconfigurable transmitarray is to integrate RF switches directly on the radiating element. Figure 5.29 shows the configuration of a transmitarray unit cell using PIN diodes [37]. This unit cell consists of a passive

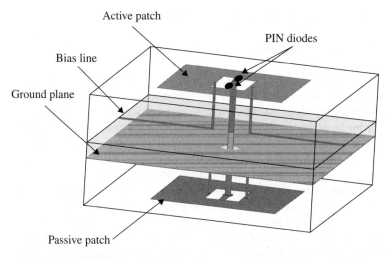

Figure 5.29 The configuration of the transmitarray unit cell using PIN diodes [37]. Reproduced with permission of ©EurAAP.

microstrip patch and an active patch. The active patch has an O-shaped slot where two PIN diodes are integrated. By alternately activating these two PIN diodes, the transmission phase of the unit cell can be controlled, operating as a 1-bit phase shifter. This configuration is similar to the unit cells used for a 1-bit reflectarray antenna design presented in section 5.3.2.

It is shown in [37] that under normal incidence the unit cell exhibits 2.1 dB of insertion loss at 9.6 GHz. If using MEMS to replace the PIN diodes, the insertion

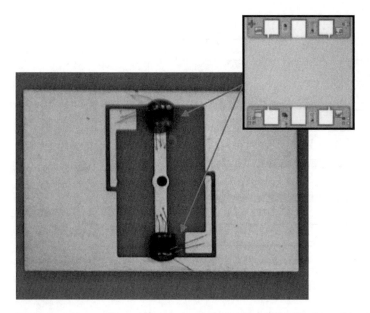

Figure 5.30 Photo of the transmitarray unit cell using MEMS [37]. Reproduced with permission of ©EurAAP.

loss can be reduced to 0.8 dB, but at higher cost. Figure 5.30 shows the fabricated photo of this transmitarray unit cell using MEMS. This element was used to design a 400-element transmitarray and the beam can be steered between ±40° with a 3 dB gain bandwidth of 14%.

Another approach to designing a beam-reconfigurable transmitarray is to introduce the RF switches on the phase delay lines of the antenna element. Because these RF components are surface mounted, there must be some air-gaps or an air-layer between the PCB substrate in order to accommodate these RF components and biasing circuitry. Thus, compared to the reflectarray, higher system complexity is expected. Figure 5.31 shows the side view of a reconfigurable transmitarray which uses a varactor to control the phase delays of the microstrip lines [38]. The unit cell is a stacked patch antenna fed by proximity coupling and is designed to resonate at 5 GHz. This configuration allows the feed lines and the radiating element to be printed at different layers, which gives more space to route the biasing circuitry and place the RF components. A similar approach is also used in the design of reconfigurable reflectarrays, as presented in the previous section. It can be seen that for a reconfigurable transmitarray design, more PCB layers are required, which increases the fabrication complexity.

As shown in Figure 5.31, the unit cell consists of one passive antenna and one active antenna. The passive antenna element is used to receive the incident waves and then transmit the received signal to the active antenna, whose microstrip feed lines are integrated with the varactors, Aeroflex GaAs MGV100-20 varactor diodes. This varactor has a capacitance tuning range of 0.15–2.0 pF. The substrate of the main radiating patch was trimmed in order to mount the varactors and DC blocking capacitors. To avoid the spurious resonances from the bias lines, the bias lines were loaded with 10 kΩ resistors so the power coupled to the bias lines can be dissipated. It is shown that

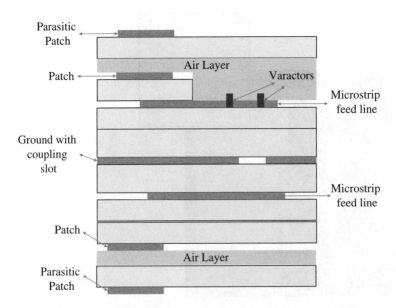

Figure 5.31 The sideview of the varactor loaded unit cell presented in [38].

the beams of the transmitarray can be scanned between ±50° with sidelobes less than −10 dB in both azimuth and elevation planes, with a gain variation of 2 dB over the 10% fractional bandwidth. The average loss of the unit cell is about 2.2 dB. This loss is mainly contributed by the dissipated loss from the parasitic resistances in the varactor diodes.

5.5 Circularly Polarised Beam-steerable Reflectarrays and Transmitarrays

Compared with linearly polarised antennas, circularly polarised antennas have several advantages, such as multi-path interference suppression, immunity of 'Faraday rotation' and reduction of polarisation mismatching. Thus, circularly polarised antennas have gained wide application, including satellite communications.

Because circular polarisation can be obtained if two orthogonal modes are excited with a 90° phase difference, it can be expressed as

$$E_{inc} = E_{inc,x}e^{j\phi} + E_{inc,y}e^{j(\phi+\pi/2)} \tag{5.13}$$

where E_{inc} is the E-field of the incident wave, and $E_{inc,x}$ and $E_{inc,y}$ are the E-field components in two orthogonal directions. For a pure circularly polarised waves, $E_{inc,x}$ and $E_{inc,y}$ have equal amplitude. If the unit cell has an orthogonally symmetrical configuration, both components experience the same phase delays

$$E_{rad} = E_{rad,x}e^{j\phi+\Delta\phi} + E_{rad,y}e^{j(\phi+\pi/2+\Delta\phi)} \tag{5.14}$$

where E_{rad} is the radiated E-field of the reflectarray or transmitarray and $\Delta\phi$ is the phase delay of the unit cell. Thus, a linearly polarised reflectarray or transmitarray can be modified to radiate circularly polarised waves by employing a circularly polarised feed. Figure 5.32 shows some examples of unit cells that have a symmetrical layout. Alternatively, with a linearly polarised feed, a circularly polarised reflectarray or transmitarray can be obtained by rotating the feed by 45° and ensuring that the re-radiated field in the x and y directions has a phase difference of 90° [39], as shown in Figure 5.33.

Sequential rotation technique is a classic methodology used in the design of circularly polarised array antennas, and this method has already been applied to the design of circularly polarised reflectarrays [40, 41]. With this approach, antenna elements are sequentially rotated by certain angles. The advantages of using this method include low cross-polarisations in the main beam, reduced mutual coupling between adjacent elements, and improved axial ratio bandwidth. However, for smart antenna design this configuration is not preferred for the following reasons:

- Limited beam-scanning capability. Since circular polarisation is obtained by sequentially rotating a group of antenna elements (e.g. four), this forms a subarray which has a very limited beam-steering capability and can result in large grating lobes when it scans the beam away from the broadside.
- Increased complexity in designing the biasing circuitry. This is also caused by the fact that the antenna elements need to be rotated as well as their phase shifts. This brings challenges to designing a biasing circuit to control the RF switches or varactors in the PCB layout.

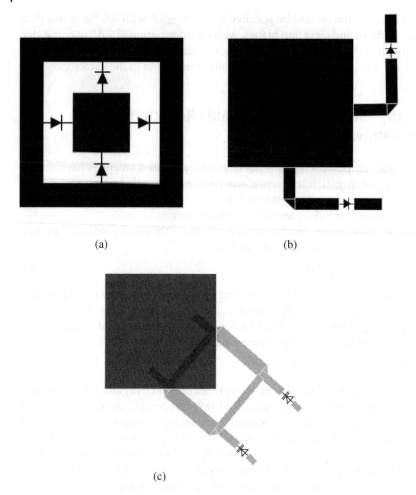

(a)   (b)

(c)

Figure 5.32 (a) The patch-loop symmetrically loaded by four varactors. (b) The dual linearly polarised patch with microstrip lines loaded by varactors. (c) The circularly polarised patch fed by a branch line coupler that is loaded by varactors.

By taking advantage of the receive–transmit architecture of the transmitarray, a transmitarray can be designed in such a way that it converts the linearly polarised incident waves into circularly polarised waves, as shown in Figure 5.34.

There are several advantages to using a linearly polarised feed. The first is that a linearly polarised feed usually has a simpler structure, is easier to design and has a lower cost. The circularly polarised feed needs to consider both the impedance and axial ratio bandwidth, in some cases they are difficult to optimise and the overlapped bandwidth is narrow. Second, most of the circularly polarised feeds have narrow beamwidth, which limits the aperture efficiency of the transmitarray or reflectarray.

Figure 5.35 shows the configuration of the unit cell of a transmitarray which transforms the incident linearly polarised wave to an outgoing circularly polarised wave [42]. The receiving antenna is a stacked rectangular patch and it receives the linearly polarised incident wave, which is then converted into the guided wave signal. The

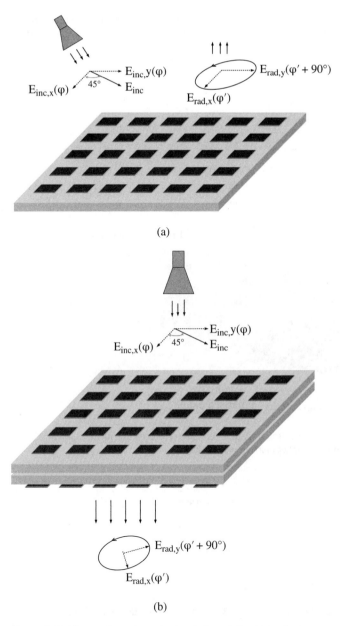

Figure 5.33 The configuration of a circularly polarised (a) reflectarray and (b) transmitarray with a linearly polarised feed rotated by 45°.

guided wave signal passes through a reflection-type phase shifter before sending it to the transmission patch antenna that is circularly polarised through a pin feed. Two varactors, M/A-COM Flip Chip MA46H120, are used to load the reflection-type phase shifter. The capacitance of this varactor can be tuned from 0.17 to 1.1 pF when the bias voltage is varied from 12 to 0 V. It is shown in [42] that about 180° continuous phase

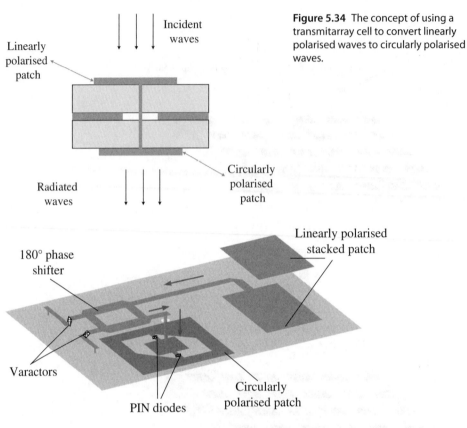

Figure 5.34 The concept of using a transmitarray cell to convert linearly polarised waves to circularly polarised waves.

Figure 5.35 The configuration of the transmitarray unit cell that transforms the incident linearly polarised wave into an outgoing circularly polarised wave [42].

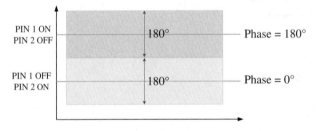

Figure 5.36 The concept of obtaining the continuous 360° phase shift range used in [42].

tuning can be obtained. The transmission antenna is a square patch with an O-shaped slot and two triangle corners along the diagonal. Two pin diodes, M/A-COM Flip Chip MA4SPS502, are integrated on the patch to realise a 1-bit differential phase shift (180°). Thus, by controlling the states of the PIN diodes and the voltages on the varactors, a continuous 360° phase shift range is obtained, as shown in Figure 5.36. Within this phase shift range, the average loss is about 2.2 dB. By tuning the transmission phase, this CP transmitarray steers its main beam within the angle range of ±45°, with a value of AR less than 3 dB and gain variation less than 2 dB.

5.6 Case Study

This case study presents a design example of a wideband multi-feed transmitarray with a reduced profile. As presented in section 5.2.2, a transmitarray antenna consists of a feed antenna and an array of unit cells. Compared to the reflectarray, the transmitarray places the feed antenna(s) directly in front of the aperture without incurring any blockage losses. This is especially advantageous when there is a need to use multiple feeds to obtain multi-beams or beam-switching. In this section, the design procedures of this wideband multi-feed transmitarray are presented.

Unit cell design   For this design example the transmitarray is based on FSSs as it aims to obtain a wideband response. The FSS works as a free-space phase shifter to obtain the desired phase delays to form the focused beam. To achieve a phase shift range of 360°, normally multiple layers of FSSs separated by a distance of a quarter-wavelength are needed [43], as shown in Figure 5.11. However, the spaces between each substrate layer lead to an increased profile of the overall antenna design. It is shown in [44] that through slightly increasing the complexity of the unit cell, a wideband transmitarray with a reduced profile can be obtained. Figure 5.37 shows the configuration of a unit cell which has three conductive layers. The unit cell consists of two pairs of tri-layer FSSs separated by a distance of a quarter-wavelength at the frequency of interest. As shown in Figure 5.37, the two square patches and one square loop with four microstrips are printed on two dielectric substrates of the same type. The tri-layer FSS is a wideband bandpass filter and the wideband response of the FSS is achieved by creating a second-order bandpass response. To achieve such a response, two capacitive layers and one inductive layer printed on two substrates are required [45].

The operation frequency of the transmitarray is chosen to be 13.5 GHz. The values of the parameters of the unit cell are listed in Table 5.1. The size of the unit cell is chosen to be approximately 0.3λ and the width of the patch is approximately 0.2λ. By resizing the unit cell, the operation frequency of the unit cell can be adjusted to other frequency bands. The chosen substrate is Roger 4003C ($\varepsilon_r = 3.55$, $\tan\sigma = 0.0027$) with thickness 0.8 mm. To better understand the operation mechanism of the FSS, the equivalent circuit model of the tri-layer FSS is shown in Figure 5.38.

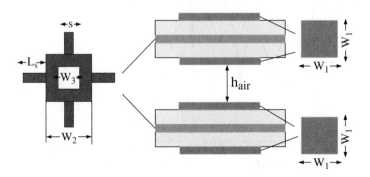

Figure 5.37 The configuration of the unit cell using two tri-layer FSSs.

Table 5.1 The values of the parameters of the unit cell.

Parameter	W_1	W_2	W_3	s	h_{air}
Value (mm)	4	3	1	0.75	4.9
Parameter	P	L_s	t_{diel}		
Value (mm)	6	1.5	0.8		

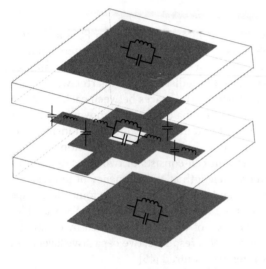

Figure 5.38 The equivalent circuit model of the tri-layer FSS working as a wideband bandpass filter.

Figure 5.39 shows the bandpass response of the unit cell. As shown, this unit cell has a wide bandpass from 11.5 to 15 GHz with low insertion loss. The phase shift of the unit cell is achieved by simultaneously varying the width (W_1) of the patches printed on the first and third layers. The size of the patches determines the upper cut-off frequency of the bandpass filter, which has a higher roll-off rate. Figure 5.40 shows the phase and

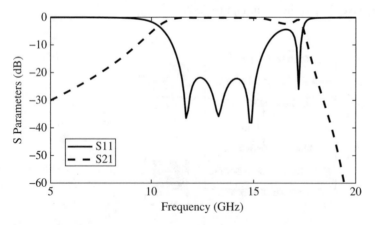

Figure 5.39 The bandpass response of the transmitarray unit cell.

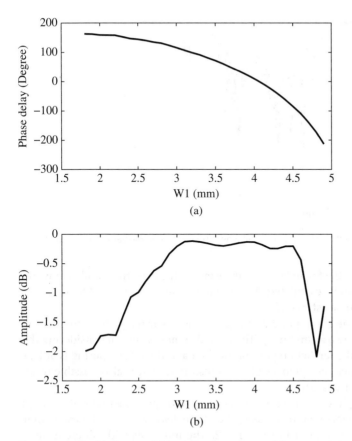

Figure 5.40 The (a) phase and (b) amplitude response of the transmission coefficient of the unit cell when the width of the patch varies at 13.5 GHz.

amplitude response of the transmission coefficient of the unit cell at 13.5 GHz when the width of the patch varies. The unit cell provides a phase shift range of 360° with insertion loss of less than 2 dB when W_1 changes from 1.8 to 4.8 mm. Within most of this region, the insertion loss is less than 0.5 dB. This feature is highly desirable in designing a high-efficiency transmitarray.

Another parameter that needs to be considered is the response of the unit cell under different incident angles. This is an important parameter because the unit cells are illuminated with waves of different incident angles. Thus, it is required that the unit cell should show a stable response regarding different incident angles. Figure 5.41 shows the reflection coefficient of the unit cell when the incident angle varies from 0° to 30°. As shown, the response of the unit cell is quite stable.

Transmitarray design After the unit cell of the transmitarray is selected, a transmitarray can be designed by following four steps:

- Decide the number of the unit cell or the aperture size of the transmitarray. The aperture size of the array antenna determines the directivity or the beamwidth of the

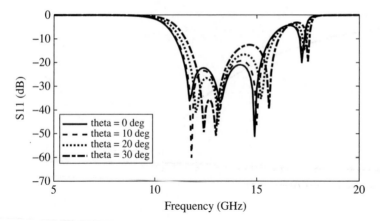

Figure 5.41 The reflection coefficient of the unit cell when the incident angle varies from 0° to 30°.

transmitarray. By using the formula of the array factor, the directivity and beamwidth of the transmitarray can be calculated. The number of unit cells needed to design the transmitarray can then be estimated.

- Choose the feed and the corresponding focal distance. Normally horn antennas are chosen as the feed of the transmitarray. The focal distance is chosen considering that the planar array is in the far-field region of the feed and the edge tapering is at least 10 dB in order to reduce the spillover of the transmitarray. It is also possible to use microstrip array antennas as the feeds.
- Calculate the desired phase distribution of the array and apply the unit cell to design the transmitarray. The formula to calculate the phase distribution of the transmitarray within the aperture is given in section 5.2. The unit cells with different phase delays need to be mapped into the aperture using the calculated phase delay curve (see Figure 5.40a) as the reference.
- Optimise the radiation performance of the transmitarray if necessary. It has been shown that the radiation performance of the reflectarray can be improved by introducing an additional reference phase when calculating the required phase for each radiating element [4, 46]. This principle can be applied to the design of the transmitarray as well

$$\phi_R = k_0(d_i - (x_i \cos \varphi_b + y_i \sin \varphi_b) \sin \theta_b) + \Delta ph \qquad (5.15)$$

where ϕ_R is the phase of the reflection coefficient of the antenna element i, k_0 is the phase constant in a vacuum, (x_i, y_i) are the coordinates of the array element i, d_i is the distance from the phase centre of the feed to the antenna unit cell, (θ_b, φ_b) are the expected scan angles of the beam in spherical coordinates, and Δph is the additional reference phase.

In this case study, the dimension of the transmitarray is chosen to be $90 \times 90 \times 8.1$ mm. There are 225 unit cells positioned in a square aperture. A linearly polarised corrugated circular horn antenna is used as the feed antenna. Considering the beamwidth of the feed, the horn antenna is placed at a distance of 90 mm above the centre of the planar

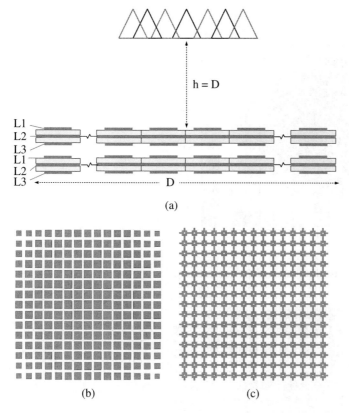

Figure 5.42 (a) Sideview of the transmitarray. (b) Layout of layers 1 and 3. (c) Layout of layer 2.

array, corresponding to a focal/diameter (f/D) ratio of 1, so the illumination tapering at the edge of the transmitarray surface is about 9 dB. Figure 5.42 shows the configuration of the designed transmitarray. As shown in Figure 5.42a, multiple feeds can be placed close to the central feed to obtain multiple beams. In this design, to optimise the performance of the transmitarray, the Δph is calculated by

$$\Delta ph = \phi' + \phi_{Max} - \phi_{Min} \tag{5.16}$$

where ϕ_{Max} and ϕ_{Min} are the maximum and minimum values of the calculated ϕ_R within the aperture of the transmitarray, respectively. ϕ' is the phase delay of the unit cell where it shows the lowest insertion loss.

Figure 5.43 shows the calculated phase distribution of the transmitarray with a central feed. The FSS elements that provide phase delays of 110° to 360° are placed in the centre of the transmitarray. Figure 5.44 shows the simulated E-field distribution on the top surface of the transmitarray. As can be seen, the incident wave from the feed antenna is concentrated on the central elements, which have very low insertion loss. The smaller size patches have slightly higher insertion loss and are placed on the edge of the planar array.

Figure 5.45 shows the radiation patterns of the transmitarray in the E plane, where multiple feeds were placed as a linear array. Because of the symmetrical configuration,

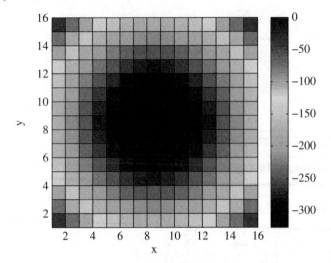

Figure 5.43 The calculated phase distribution of the transmitarray.

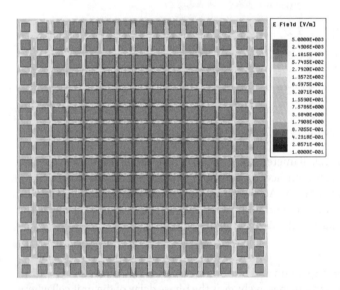

Figure 5.44 The E-field distribution on the top surface of the transmitarray.

the radiation patterns are very symmetrical. It can be seen that the beams of the transmitarray can be switched from −15° to 15° with gain variation of 2 dB. The gain reduction is mainly caused by the limited aperture size of the transmitarray and by offsetting the feed; the aperture efficiency of the transmitarray is reduced due to increased spillover.

Figure 5.46 shows the measured gain variation of the designed transmitarray within its operating frequency band for broadside radiation. It is shown that from 12 to 14.5 GHz the gain variation is within 1.5 dB, representing a fractional bandwidth of 18.5%. The 1 dB gain variation bandwidth is 16%.

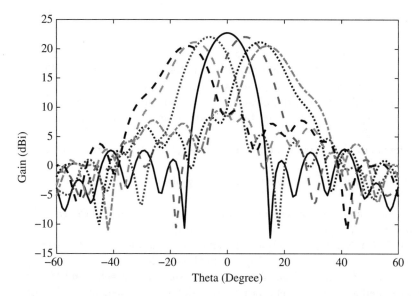

Figure 5.45 The radiation patterns of the transmitarray in the E plane with multiple feeds.

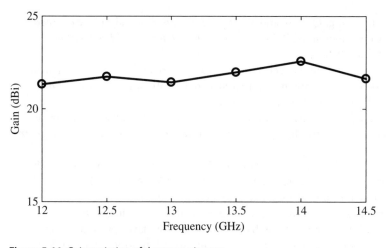

Figure 5.46 Gain variation of the transmitarray.

5.7 Summary of the Chapter

In this chapter the use of reflectarrays and transmitarrays to design beam-reconfigurable array antennas is presented. The techniques to realise multi-beam or beam-switching include the use of multiple feeds and incorporating RF switches or varactors on the antenna elements. Both types of array antenna are suitable for high-gain applications while avoiding the use of a feed network, thus they are good candidates for low-cost smart antennas. Meanwhile, they have great potential to be used in emerging 5G mm-wave communications.

References

1 M. Jiang, Z.N. Chen, Y. Zhang, W. Hong, and X. Xuan. Metamaterial-based thin planar lens antenna for spatial beamforming and multibeam massive mimo. *IEEE Transactions on Antennas and Propagation*, 65(2):464–472, Feb 2017.

2 S. Foo. Metamaterial-based transmitarray for orthogonal-beam-space massive-mimo. In *2016 10th European Conference on Antennas and Propagation (EuCAP)*, pages 1–5, April 2016.

3 S. Foo. Metamaterial-based transmitarray for multi-beam antenna array assemblies, 2017.

4 J. Huang; J.A. Encinar. *Reflectarray Antennas*. Wiley-Blackwell, Hoboken, New Jersey, 2007.

5 P. Nayeri, F. Yang, and A.Z. Elsherbeni. *Reflectarray Antennas: Theory, Designs, and Applications*. John Wiley & Sons Ltd, Chichester, 2018.

6 A.H. Abdelrahman, F. Yang, A.Z. Elsherbeni, and P. Nayeri. *Analysis and Design of Transmitarray Antennas*. Morgan & Claypool Publishers, 2017.

7 Q. Luo, S. Gao, C. Zhang, D. Zhou, T. Chaloun, W. Menzel, V. Ziegler, and M. Sobhy. Design and analysis of a reflectarray using slot antenna elements for Ka-band SatCom. *IEEE Transactions on Antennas and Propagation*, 63(4):1365–1374, April 2015.

8 W. Li, S. Gao, L. Zhang, Q. Luo, and Y. Cai. An ultra-wide-band tightly coupled dipole reflectarray antenna. *IEEE Transactions on Antennas and Propagation*, 66(2):533–540, Feb 2018.

9 C. Zhang, Y. Wang, F. Zhu, G. Wei, J. Li, C. Wu, S. Gao, and H. Liu. A planar integrated folded reflectarray antenna with circular polarization. *IEEE Transactions on Antennas and Propagation*, 65(1):385–390, Jan 2017.

10 W. Menzel, D. Pilz, and M. Al-Tikriti. Millimeter-wave folded reflector antennas with high gain, low loss, and low profile. *IEEE Antennas and Propagation Magazine*, 44(3):24–29, 2002.

11 D.M. Pozar and T.A. Metzler. Analysis of a reflectarray antenna using microstrip patches of variable size. *Electronics Letters*, 29(8):657–658, April 1993.

12 J.A. Encinar. Design of a dual frequency reflectarray using microstrip stacked patches of variable size. *Electronics Letters*, 32(12):1049–1050, Jun 1996.

13 J.A. Encinar and J.A. Zornoza. Broadband design of three-layer printed reflectarrays. *IEEE Transactions on Antennas and Propagation*, 51(7):1662–1664, July 2003.

14 M.Y. Ismail, W. Hu, R. Cahill, V.F. Fusco, H.S. Gamble, D. Linton, R. Dickie, S.P. Rea, and N. Grant. Phase agile reflectarray cells based on liquid crystals. *IET Microwaves, Antennas Propagation*, 1(4):809–814, Aug 2007.

15 J. Huang and R.J. Pogorzelski. A Ka-band microstrip reflectarray with elements having variable rotation angles. *IEEE Transactions on Antennas and Propagation*, 46(5):650–656, May 1998.

16 E. Almajali, D.A. McNamara, N. Gagnon, and A. Petosa. Remarks on the feasibility of full-wave analyses of printed lens/transmitarray antennas. In *2014 IEEE Antennas and Propagation Society International Symposium (APSURSI)*, pages 1268–1269, July 2014.

17 E. Plaza, G. Leon, S. Loredo, and F. Las-Heras. A simple model for analyzing transmitarray lenses. *IEEE Antennas and Propagation Magazine*, 57(2):131–144, April 2015.

18 J.Y. Lau and S.V. Hum. A wideband reconfigurable transmitarray element. *IEEE Transactions on Antennas and Propagation*, 60(3):1303–1311, March 2012.

19 H. Kaouach, L. Dussopt, J. Lanteri, T. Koleck, and R. Sauleau. Wideband low-loss linear and circular polarization transmit-arrays in V-band. *IEEE Transactions on Antennas and Propagation*, 59(7):2513–2523, July 2011.

20 H. Nematollahi, J.J. Laurin, J.E. Page, and J.A. Encinar. Design of broadband transmitarray unit cells with comparative study of different numbers of layers. *IEEE Transactions on Antennas and Propagation*, 63(4):1473–1481, April 2015.

21 C.G.M. Ryan, M.R. Chaharmir, J. Shaker, J.R. Bray, Y.M.M. Antar, and A. Ittipiboon. A wideband transmitarray using dual-resonant double square rings. *IEEE Transactions on Antennas and Propagation*, 58(5):1486–1493, May 2010.

22 W. Menzel, M. Al-Tikriti, and R. Leberer. A 76 GHz multiple-beam planar reflector antenna. In *2002 32nd European Microwave Conference*, pages 1–4, Sept 2002.

23 W. Menzel and D. Kessler. A folded reflectarray antenna for 2D scanning. In *2009 German Microwave Conference*, pages 1–4, March 2009.

24 M. Arrebola, J.A. Encinar, and M. Barba. Multifed printed reflectarray with three simultaneous shaped beams for LMDS central station antenna. *IEEE Transactions on Antennas and Propagation*, 56(6):1518–1527, June 2008.

25 O.M. Bucci, G. Franceschetti, G. Mazzarella, and G. Panariello. Intersection approach to array pattern synthesis. *IEE Proceedings H – Microwaves, Antennas and Propagation*, 137(6):349–357, Dec 1990.

26 R.J. Mailloux. *Phased Array Antenna Handbook*. Artech House Inc., Norwood, MA, 2015.

27 H. Kamoda, T. Iwasaki, J. Tsumochi, T. Kuki, and O. Hashimoto. 60-GHz electronically reconfigurable large reflectarray using single-bit phase shifters. *IEEE Transactions on Antennas and Propagation*, 59(7):2524–2531, July 2011.

28 M.T. Zhang, S. Gao, Y.C. Jiao, J.X. Wan, B.N. Tian, C.B. Wu, and A.J. Farrall. Design of novel reconfigurable reflectarrays with single-bit phase resolution for Ku-band satellite antenna applications. *IEEE Transactions on Antennas and Propagation*, 64(5):1634–1641, May 2016.

29 O. Bayraktar, O.A. Civi, and T. Akin. Beam switching reflectarray monolithically integrated with RF MEMS switches. *IEEE Transactions on Antennas and Propagation*, 60(2):854–862, Feb 2012.

30 E. Carrasco, M. Barba, and J.A. Encinar. X-band reflectarray antenna with switching-beam using pin diodes and gathered elements. *IEEE Transactions on Antennas and Propagation*, 60(12):5700–5708, Dec 2012.

31 E. Carrasco, M. Arrebola, M. Barba, and J.A. Encinar. Shaped-beam reconfigurable reflectarray with gathered elements in an irregular lattice for LMDS base station. In *Proceedings of the 5th European Conference on Antennas and Propagation (EUCAP)*, pages 975–978, April 2011.

32 J. Perruisseau-Carrier and A.K. Skrivervik. Monolithic MEMS-based reflectarray cell digitally reconfigurable over a 360° phase range. *IEEE Antennas and Wireless Propagation Letters*, 7:138–141, 2008.

33 Skyworks. *Varactor Diodes: Application Notes*, 8, 2018. 200824Rev.

34 F. Venneri, S. Costanzo, G. Di Massa, A. Borgia, P. Corsonello, and M. Salzano. Design of a reconfigurable reflectarray based on a varactor tuned element. In *2012 6th European Conference on Antennas and Propagation (EUCAP)*, pages 2628–2631, March 2012.

35 A. Tayebi, J. Tang, P.R. Paladhi, L. Udpa, S.S. Udpa, and E.J. Rothwell. Dynamic beam shaping using a dual-band electronically tunable reflectarray antenna. *IEEE Transactions on Antennas and Propagation*, 63(10):4534–4539, Oct 2015.

36 M. Zhang, S. Gao, C. Gu, J. Wan, and C. Wu. Varactor-loaded dual-polarized unit-cell for reconfigurable reflectarrays. In *8th European Conference on Antennas and Propagation (EuCAP 2014)*, pages 1950–1954, April 2014.

37 A. Clemente, L. Dussopt, B. Reig, R. Sauleau, P. Potier, and P. Pouliguen. Reconfigurable unit-cells for beam-scanning transmitarrays in X band. In *2013 7th European Conference on Antennas and Propagation (EuCAP)*, pages 1783–1787, April 2013.

38 J.Y. Lau and S.V. Hum. Reconfigurable transmitarray design approaches for beamforming applications. *IEEE Transactions on Antennas and Propagation*, 60(12):5679–5689, Dec 2012.

39 M.R. Chaharmir, J. Shaker, M. Cuhaci, and A. Sebak. Circularly polarised reflectarray with cross-slot of varying arms on ground plane. *Electronics Letters*, 38(24):1492–1493, Nov 2002.

40 A.E. Martynyuk, J.I.M. Lopez, and N.A. Martynyuk. Spiraphase-type reflectarrays based on loaded ring slot resonators. *IEEE Transactions on Antennas and Propagation*, 52(1):142–153, Jan 2004.

41 B. Strassner, C. Han, and Kai Chang. Circularly polarized reflectarray with microstrip ring elements having variable rotation angles. *IEEE Transactions on Antennas and Propagation*, 52(4):1122–1125, April 2004.

42 C. Huang, W. Pan, and X. Luo. Low-loss circularly polarized transmitarray for beam steering application. *IEEE Transactions on Antennas and Propagation*, 64(10):4471–4476, Oct 2016.

43 A.H. Abdelrahman and et al. Bandwidth improvement methods of transmitarray antennas. *IEEE Transactions on Antennas and Propagation*, 63(7):2946–2954, July 2015.

44 Q. Luo, S. Gao, M. Sobhy, and X. Yang. Wideband transmitarray with reduced profile. *IEEE Antennas and Wireless Propagation Letters*, 17(3):450–453, March 2018.

45 M.A. Al-Joumayly and N. Behdad. A generalized method for synthesizing low-profile, band-pass frequency selective surfaces with non-resonant constituting elements. *IEEE Transactions on Antennas and Propagation*, 58(12):4033–4041, Dec 2010.

46 Y. Mao, S. Xu, F. Yang, and A.Z. Elsherbeni. A novel phase synthesis approach for wideband reflectarray design. *IEEE Transactions on Antennas and Propagation*, 63(9):4189–4193, Sept 2015.

6

Compact MIMO Antenna Systems

6.1 Introduction

MIMO technology uses multiple antennas at the transmitter and receiver to increase the channel throughput and achieve high data rate wireless communications by taking advantage of the space diversity. According to the Shannon–Hartley theorem, the maximum data rate where information can be transmitted over a communications channel is

$$C = B\log_2\left(1 + \frac{S}{N}\right) \tag{6.1}$$

where C is the channel capacity in bits per second, B is the bandwidth of the channel in hertz, and S/N is the SNR. Considering a MIMO system with fixed linear $n \times n$ matrix channel (n transmit and n receive antennas) with additive white Gaussian noise, the MIMO capacity can be expressed as [1]

$$C = n \cdot \log_2\left(1 + \frac{\rho}{n}\right) + \log_2\Delta_n \tag{6.2}$$

where ρ is the average SNR and Δ_n is related to the correlation matrix of the channel. As shown, MIMO capacity scales linearly as the number of antennas if the multipath channels are not correlated. Figure 6.1 shows the system diagram of a MIMO system.

MIMO technology has been included in several wireless standards, including wireless LAN, WiMAX, long-term evolution (LTE), and fifth-generation (5G) mobile networks. For example, in 802.11n, it is defined that the MIMO antennas should operate in both the 2.4 GHz and the 5 GHz bands, with a maximum net data rate from 54 to 600 Mbit/s [2]. In the 5G standard, mm-wave MIMO antennas are required to be used on either the mobile terminals and base stations, and are expected to achieve data rates of tens of megabits per second for tens of thousands of users by using a massive MIMO system [3].

6.2 MIMO Antennas

The MIMO antenna is a critical component of a MIMO system as it determines the performance of the system, including the data rate and channel bandwidth. When designing an array antenna for MIMO applications several key figures of merit should be considered.

Low-cost Smart Antennas, First Edition. Qi Luo, Steven (Shichang) Gao, Wei Liu, and Chao Gu.
© 2019 John Wiley & Sons Ltd. Published 2019 by John Wiley & Sons Ltd.

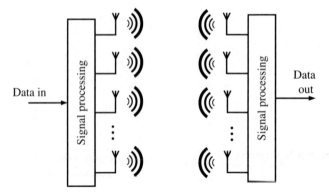

Figure 6.1 The MIMO system.

6.2.1 Isolation

Isolation between antennas is a measure of how much power is coupled from one antenna element to other antenna elements in a multiple antenna system. It is measured by using a vector network analyser and checking the transmission coefficients (S21 and S12). For MIMO systems, it is highly desirable to have high isolation between the antenna elements, otherwise the system suffers low efficiency because some of the input power is coupled to other radiating elements instead of radiating to the free space. One approach to increase the isolation is to increase the physical separation between the radiating elements. Figure 6.2 shows the simulated isolation between two antenna elements in the H-plane with varied distance. The antenna element is a stacked square patch resonating at 9.9 GHz. As shown, when the distance between the two antenna elements increases from 0.5λ to λ, where λ is the wavelength in free space, the isolation increases from 16.6 to 30.5 dB.

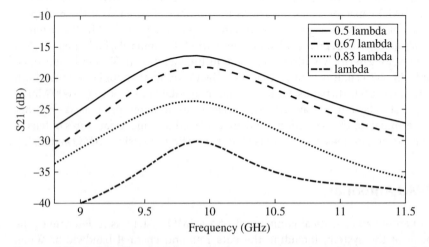

Figure 6.2 The isolation between two microstrip patch antennas in the H plane with varied distance.

However, increasing the distance between the antenna elements is not suitable for compact size devices where there is limited space to accommodate the antennas. Many techniques are reported to overcome this problem, including the use of decoupling networks [4–6], neutralisation lines [7–9], and defect ground systems (DGS) [10]. These techniques are discussed in next section.

6.2.2 Envelope Correlation Coefficient

The envelope correlation coefficient (ECC) is an important parameter to measure the diversity gain of a MIMO system. It is a measure of the correlation between the radiation of the antenna elements and the communication channels. The value of the ECC ranges from 0 to 1, where 0 represents no correlation and 1 represents completely correlated channels. Low envelope correlation means high diversity gain. The envelope correlation between two antennas is [11]

$$\rho_e = \frac{|\iint_{4\pi} [\vec{F}_1(\theta, \phi) \cdot \vec{F}_2(\theta, \phi)] d\Omega|^2}{\iint_{4\pi} |\vec{F}_1(\theta, \phi) d\Omega|^2 \cdot \iint_{4\pi} |\vec{F}_2(\theta, \phi) d\Omega|^2} \tag{6.3}$$

where $\vec{F}_i(\theta, \phi)$ is the far-field radiation pattern of the antenna element when port i is excited and $\cdot$ presents the Hermitian product. Normally, measurement of the radiation pattern is needed to calculate this parameter. However, if the antenna shows high efficiency, a simple formula which only uses the scattering matrix is derived to calculate the ECC [11]

$$|\rho_e| = \frac{|S_{11}^* S_{12} + S_{21}^* S_{22}|^2}{(1 - (|S_{11}|^2 + |S_{21}|^2))(1 - (|S_{22}|^2 + |S_{12}|^2))} \tag{6.4}$$

Figure 6.3 shows the calculated ECC of two antennas spaced by 0.5λ using Equation 6.4. The antenna element is the same square patch used in Figure 6.2, which

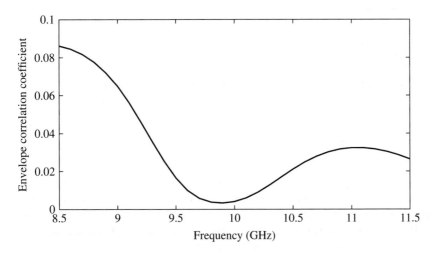

Figure 6.3 The calculated ECC of two antennas.

resonates at 9.9 GHz. The 10 dB return loss bandwidth of the radiating element is from 9.7 to 10.1 GHz. At 10 GHz, the isolation is about 17 dB. As shown, within its bandwidth the ECC is close to zero, indicating that using these antennas in the MIMO system can achieve high diversity gain.

Equation 6.4 is based on the assumption that the antennas are lossless. This is not true in practical cases since there are always losses associated with one antenna, such as conduction loss and antenna mismatching loss. In order to refine this method, the radiation efficiency is included in the calculation of the correlation coefficient and by doing this the uncertainty of the calculated correlation coefficient can be derived [12]

$$Uncertainty = \pm \sqrt{(\frac{1}{\eta_1} - 1)(\frac{1}{\eta_2} - 1)} \tag{6.5}$$

where η_1 and η_2 are the radiation efficiencies of the two antennas, respectively.

6.2.3 Total Active Reflection Coefficient

The total active reflection coefficient (TARC) was firstly introduced by [13] to describe the properties of the multi-port antennas. The TARC characterises the bandwidth of an N-port antenna, which shows the true bandwidth of the MIMO antenna. It is a real number between zero and one. If all of the incident power is radiated, the value of the TARC is equal to zero. On the other hand, if the value of TARC is one, it implies that all of the incident power is either reflected back or coupled to other ports of the MIMO antenna. The TARC is defined as the square root of the reflected power divided by the incident power [13]

$$\Gamma_a^t = \sqrt{\frac{p_a - p_r}{p_a}} \tag{6.6}$$

where Γ_a^t is the TARC, p_a is the incident power, and p_r is the radiated power. If an N-port antenna is excited at the ith port and the other ports are connected to matched loads, the TARC can be calculated by

$$\Gamma_a^t = \sqrt{\sum_{j=1}^{N} |S_{ij}|^2}, i = 1, ...N \tag{6.7}$$

where S_{ij} is the scattering parameters. By assuming the reflected signal is unity magnitude randomly phased with Gaussian random variable, the TARC of a two-port MIMO antenna is described as

$$\Gamma_a^t = \frac{\sqrt{(|S_{11} + S_{12}e^{j\theta}|^2) + (|S_{21} + S_{22}e^{j\theta}|^2)}}{\sqrt{2}} \tag{6.8}$$

where θ is the phase of the input signal. Using the same two-antenna array as discussed for the ECC, Figure 6.4 compares the calculated TARC with the reflection coefficient of the isolated antenna element. Generally speaking, the TARC has a narrower bandwidth than the impedance bandwidth of the antenna element, and it is determined by the isolation and active return loss of the antennas.

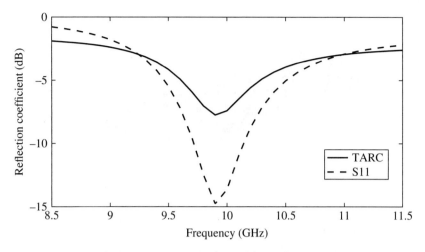

Figure 6.4 Comparison between the calculated TARC and the reflection coefficient of the antenna element.

6.3 Compact MIMO Antenna with High Isolation

The biggest challenge to design compact MIMO array antennas is how to obtain high isolation between closely spaced antenna elements while maintaining good radiation efficiency of the antenna. To have good space diversity, traditionally the space between each antenna element is required to be approximately half of the wavelength. However, for most portable wireless devices it is impossible to follow this design rule due to the limited available volumes. In this section, some practical solutions to design compact antenna arrays are presented.

6.3.1 Neutralisation Technique

The neutralisation technique was first proposed to improve the isolation of two closely spaced planar inverted-F antennas (PIFAs) [14]. Using this method, the isolation between two antennas at certain frequencies is improved through neutralising the current of two antennas or producing opposite couplings without the need to add extra space for antenna design. According to [15], the location of the neutralising line needs to be placed where the surface current is maximum (minimum E-field) and the length of it needs to be approximately a quarter-wavelength.

Besides applying to PIFAs, the neutralisation technique can also be applied to other types of antennas, such as printed monopoles for 4G mobile terminals [7, 16, 17] and WLAN USB dongles [18, 19]. In [18], a compact and low-cost inverted-L antenna (ILA) array was developed for the MIMO WLAN application at 5.8GHz (IEEE 802.11n). The ILA is a bent monopole antenna and the total length of the inverted-L needs to be approximately one-quarter of the wavelength at the resonant frequency of interest. However, the challenge for this work is that the two antennas need to be closely located

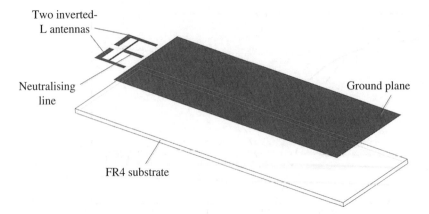

Two inverted-
L antennas

Neutralising
line

Ground plane

FR4 substrate

Figure 6.5 The configuration of the compact ILA array for USB dongle application [21].

in a small area of a USB dongle, where the available size on the PCB for printing the antennas is less than 10 mm × 17 mm (approximately $0.2\lambda_{5.8GHz} \times 0.3\lambda_{5.8GHz}$). Figure 6.5 shows the configuration of the compact ILAs array presented in [18]. This antenna array has two equal ILAs symmetrically mirrored to each other. A neutralising line is added between these two antenna elements to improve the isolation. Normally, the ILA antenna has a low input impedance [20]. The typical method employed to solve this problem for an ILA is to short the antenna element to the ground plane and change the feeding position, which in turn increases the input impedance of the antenna. However, shorting the antenna to the ground plane increases the impact of the ground plane size to the radiation performance of the antenna. Instead, the technique proposed in [18] improves the impedance matching of the antenna array by including one vertical stub in the middle of the neutralising line, as depicted in Figure 6.5. From the aspect of the antenna array, adding this stub has little influence on the isolation between the two antennas as the isolation is mainly controlled by the length, width, and position of the horizontal neutralizing line. Meanwhile, for the single antenna itself, the equivalent antenna structure is one bent monopole with an L-shaped stub, which operates as an impedance transformer. Figure 6.6 shows the measured scattering parameters of this compact ILA array. The measurement results suggest that the proposed ILA array has a 10 dB return loss bandwidth from 5.7 to more than 6 GHz. Within the desired WLAN operation band, isolation of better than 12 dB is obtained.

Another printed 2.4 GHz MIMO WLAN antenna incorporating a neutralisation line for antenna port decoupling is presented in [19]. The two monopoles are located on the two opposite corners of the PCB, which is 30 × 65 mm. After adding the neutralisation line, the antenna port isolation is increased by 9 dB compared to the case where there is no neutralisation line presented. The measurement results show that the antenna element has a peak gain of 2.1 dBi with radiation efficiency more than 70%, which means that introducing the neutralisation line does not affect the radiation efficiency of the antenna. To obtain dual-band operation, the concept of introducing RF switches on the neutralisation line is presented in [9]. In this work, two RF switches were symmetrically embedded on the neutralisation line of a compact C-shaped printed two monopole antenna array, as shown in Figure 6.7. By switching the states of the RF switches, the array can be reconfigured to have high isolation at WLAN 2.4 or 5.2 GHz bands.

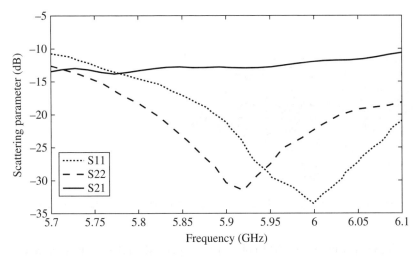

Figure 6.6 The measured scattering parameters of the compact ILA array [21].

Figure 6.7 The concept of introducing RF switches on a neutralising line to realise the dual-band operation represented in [9].

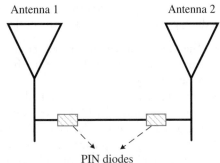

One constraint of the neutralisation technique is its limited bandwidth. To improve the bandwidth, a wideband neutralisation line was presented in [22]. This modified neutralisation line consists of two metal strips and a metal circular disc. Figure 6.8 shows the configuration of this design. The distance between the centres of the monopole antennas is only 9.6 mm ($0.13\lambda_{4GHz}$). The wideband operation is obtained because the circular disc allows multiple decoupling current paths with different lengths to cancel the coupling currents. The experimental results given in [22] show that the isolation from 3.1 to 5 GHz is increased to higher than 22 dB after introducing the wideband neutralisation line while the ECC is lower than 0.1 across the operating band.

6.3.2 Metamaterial

MTMs, which are artificial structures that have many properties not found in nature, have attracted significant attention over recent years. MTM structures have been applied in various ways to antenna designs, including antenna minimisation [23, 24], antenna radiation performance enhancement [25–27], and suppressing the mutual coupling of antenna elements in array antennas [28, 29].

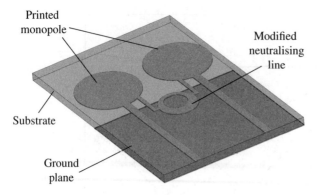

Printed monopole

Modified neutralising line

Substrate

Ground plane

Figure 6.8 The configuration of the monopole array with the wideband neutralisation line reported in [22].

One approach to applying the MTM to obtain a compact MIMO array is to place the planar MTM between the antenna elements, as shown in Figure 6.9. The planar MTM is preferred for this scenario because planar configuration means that it can be printed on the same layer of the microstrip antenna, thus simplifying the fabrication process and reducing the fabrication cost. The design guidelines are as follows:

1. The MTM unit cell must have compact size, for instance of subwavelength dimensions, otherwise it will invalidate the design objective, which is to obtain a compact array antenna.
2. The MTM needs to be designed to block or suppress the surface wave propagation between the MIMO antenna elements. The S parameters of the MTM can be extracted by EM simulations with periodic boundary conditions.
3. After placing the planar MTM between the MIMO antenna elements, it is important to ensure that the impedance matching of the antennas is not affected. The MTMs inevitably introduce additional capacitive coupling to the radiating elements, thus the location of the MTM must be optimised. In addition, the feed point of the antenna can be adjusted to compensate the additional capacitance contributed from the MTM.
4. During the simulation, the MTM unit cell is simulated with the periodic boundary condition. However, there is a finite number of MTM cells placed between the antenna elements. Therefore, the dimension of the MTM cell needs to be adjusted to ensure it operates at the desired frequencies.

A good example of this approach is to use a split-ring resonator (SRR) to improve the isolation of a two-element printed MIMO monopole array antenna [30]. In this study, the single SRR unit cell is placed at different locations between the two monopoles, as shown in Figure 6.10. The monopole has a length of a quarter-wavelength and was designed to resonate at 2.4 GHz. The distance between the two antennas is 14 mm, corresponding to $0.11\lambda_{2.4GHz}$. This dimension of the SRR is in the order of $\lambda/10$. The rationale of where the SRR should be placed is to study the propagation paths of the EM fields that are responsible for the mutual coupling. It is identified in [30] that there are two areas with strong EM fields, one close to the ground plane and the other near the tips of the monopole.

Figure 6.9 Placing planar MTM between two antennas to reduce the mutual couplings.

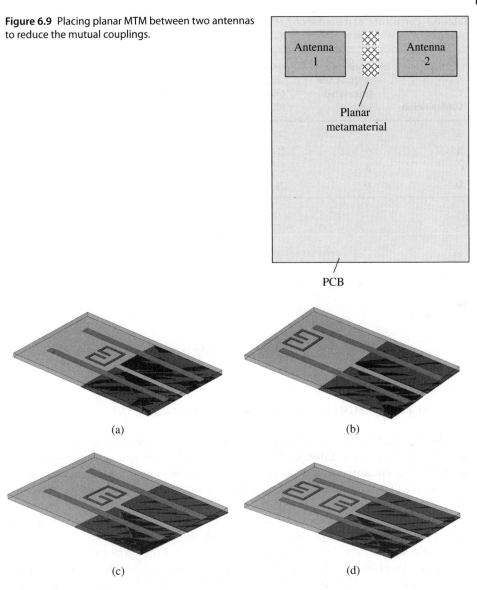

Figure 6.10 Different configurations of SRR placement between the two monopole array antennas.

The measurement results from [30] show that with all of the four configurations presented in Figure 6.10 the isolation between the two monopoles is increased from 6 dB to more than 20 dB, while there is little effect on the radiation patterns and radiation efficiency of the array antenna. Wider bandwidth is obtained when there are two resonators placed between the monopole elements. Table 6.1 summarises the radiation performance of these MIMO arrays.

Besides placing the planar MTM on the same layer of the antenna elements, it can also be used as the isolation wall between closely spaced array antennas, as shown in Figure 6.11.

Table 6.1 Comparison of the radiation performances of the monopole array with SRR at different locations.

Configuration	Isolation without SRR (measured) (dB)	Isolation with SRR (measured) (dB)	Measured radiation efficiency (%)
A	6	20	60
B	6	18	65
C	6	23	71
D	6	24	70

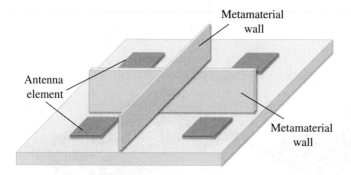

Figure 6.11 Placing an MTM wall in a planar array to improve isolation.

A double-layer mushroom structure is used to enhance the inter-element isolation of a four-element array antenna [31]. The antenna element is a substrate-integrated cavity-backed slot antenna and the distance between the antenna elements is approximately a half-wavelength. The double-layer mushroom is designed to have the surface-wave forbidden band between 2.4 and 2.8 GHz. After introducing the MTM wall, the isolation between the antenna elements increases to more than 42 dB and the ECC is lower than 0.02 across the operating bandwidth.

Recently, studies have shown that besides making use of the surface wave suppressing characteristics of the MTM, the mutual coupling between the MIMO array can also be reduced by using a MTM polarisation-rotator (MPR) wall [29]. By embedding an MPR wall which consists of a 1 × 7 unit cell between two dielectric resonator antennas that resonate at 60 GHz, the mutual coupling is reduced by more than 16 dB on average. The MPR wall converts the TE modes of the antennas to orthogonal polarisation, and effectively reduces the mutual coupling from the spatial field of the antennas without affecting their radiation patterns. Figure 6.12 shows the configuration of the MPR unit cell. As shown, the MPR has three layers: an SRR on the first and third layers, and a coupling strip in the middle layer. According to [29], the radii of the SRR controls the unit cell resonance frequency and the metallic strip in the middle is used to rotate the E-field.

Figure 6.12 Configuration of the MPR unit cell developed in [29].

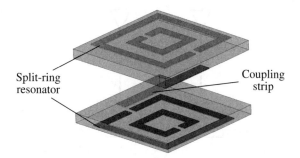

Split-ring resonator

Coupling strip

Compare to planar MTMs, the MTM wall is more effective in reducing mutual couplings between the closely spaced array antennas. The trade-off is that it increases the fabrication complexity and requires additional volume.

6.3.3 Decoupling Network

The principle of the decoupling network is to introduce additional circuitry to minimise the couplings between the antenna elements by either cancelling the couplings or trapping the coupling currents. The mutual coupling between two closely spaced antennas can be represented using a 2×2 admittance matrix [5]

$$Y = \begin{bmatrix} Y_{11}^A & Y_{12}^A \\ Y_{21}^A & Y_{22}^A \end{bmatrix} \tag{6.9}$$

where $Y_{i,j}^A$ is the mutual admittance of the coupled antennas.

The mutual coupling of the two antennas can be cancelled by introducing a second-order or higher order coupled resonator network that is connected to the two coupled antennas in shunt. Then the mutual coupling of two closely spaced antennas can be written as

$$Y = \begin{bmatrix} Y_{11}^A + Y_{11}^N & Y_{12}^A + Y_{12}^N \\ Y_{21}^A + Y_{21}^N & Y_{22}^A + Y_{22}^N \end{bmatrix} \tag{6.10}$$

where $Y_{i,j}^N$ is the admittance matrix of the decoupling network. If the following condition can be satisfied, high isolation between the two ports of the connected network is obtained

$$Y_{21}(\omega_r) = Y_{21}(\omega_r)^A + Y_{21}(\omega_r)^N = 0 \tag{6.11}$$

where ω_r is the resonant angular frequency of the antennas. Thus, in theory, the isolation between antennas can be improved by introducing a well-designed external decoupling network. The decoupling network can be realised by either lumped elements or microstrips. Matching networks are usually required, especially for broadband or dual-band designs. A fully lumped decoupling network was presented in [32]. The decoupling network comprises two series and four shunt reactive elements. Figure 6.13 shows the circuit of this decoupling network. In the even mode, the half-circuit of the

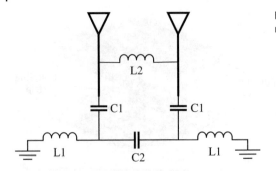

Figure 6.13 The lumped decoupling network presented in [32].

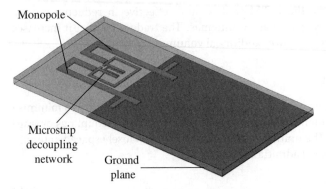

Figure 6.14 The configuration of the compact monopole array with a microstrip decoupling network presented in [33].

decoupling network is identical to an L-section matching network. In the odd mode, it is equivalent to a π section filter. In this way, the compact size decoupling network simultaneously achieves good impedance matching and high port isolation of the two element antenna array. From the results shown in [32], this decoupling network is able to be fitted in a small area, where the centre-to-centre spacing between the two printed monopoles is about 0.075λ. After introducing this lumped decoupling network, the ECC reduced from 0.51 to 0.08 at the resonance (2.45 GHz) of the array antenna, the peak gain increased from −0.8 to 2.1 dBi, and the radiation efficiency increased from 44.9% to 70.3%.

Figure 6.14 shows the configuration of a compact monopole array that uses microstrips to design the decoupling network [33]. The distance between the two monopoles is less than 0.01λ. The decoupling network is realised by two meander lines which form the distributed capacitors and inductors. The inductive coupling is obtained by a connecting line while the capacitive coupling is realised by using two gaps between a π-shaped line and two meander stubs. By optimising the dimensions of the meander lines, including the length of the connecting line and the width of the π shaped microstrip line, the couplings between the two monopoles can be cancelled out, which leads to increased isolation between the antenna ports. Figure 6.15 compares the ECC and radiation efficiency of the array antenna with and without the decoupling network. As shown, the total efficiency of the array is improved by over 10% after the decoupling network is introduced.

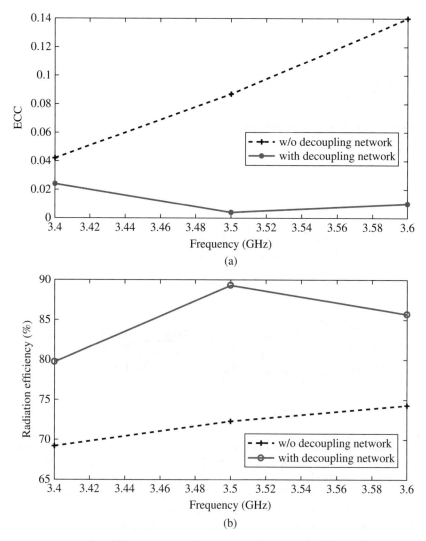

Figure 6.15 Comparison of the ECC and radiation efficiency of the array antenna reported in [33] with and without the decoupling network.

In the scenario where a wide isolation bandwidth must be obtained between closely spaced antennas, e.g. for 5G wireless systems, it is necessary to use multiple decoupling elements that resonate at different frequencies to achieve multimode decoupling. For a mobile terminal, the available space to accommodate the antennas is very limited. This means the decoupling elements are also closely spaced, which leads to couplings between the decoupling elements. This coupling effect can affect the function of the decoupling element due to the mutual couplings. To address this issue, a multimode decoupling element was reported [34]. Although multimode decoupling elements can be used to replace a number of single-mode decoupling elements, thus alleviating the mutual coupling between the decoupling elements, the multimode decoupling element is difficult to tune and increases the fabrication complexity. Recently, a new approach

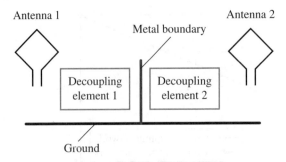

Antenna 1

Metal boundary

Antenna 2

Figure 6.16 The concept of introducing a metal boundary between the two closely space decoupling elements presented in [35].

Decoupling element 1

Decoupling element 2

Ground

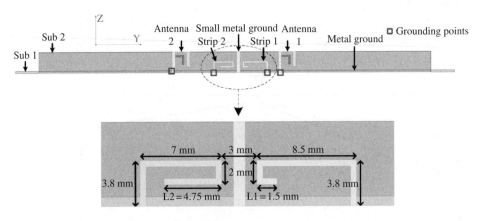

Figure 6.17 The configuration of the side-edge antenna array for smartphones [35].

that can solve the problem of a strong mutual effect between the single-mode coupling elements was presented [35]. This approach introduces a metal boundary between the two closely space decoupling elements and thus the operating modes of different decoupling elements can be independently tuned even if the edge-to-edge distance between these decoupling elements is less than $0.04\lambda_0$. Figure 6.16 illustrates this concept.

To demonstrate the effectiveness of this approach, a side-edge antenna array for smartphone application was designed [35] and the configuration of the array antenna is shown in Figure 6.17. The antenna elements are shorted strips that are fed by coupling lines. There are three PCBs, Sub 1, Sub 2, and Sub 3, all of which are 0.8 mm thick, and a double-sided FR4 substrate ($\varepsilon_r = 4.4$ and loss tangent $= 0.02$). Sub 2 and Sub 3 (134 × 6.2 × 0.8 mm³ each) are perpendicularly placed on top of Sub 1. The antenna elements are chosen to resonate at 3.5 GHz. The dimension of the array is 150 × 75 × 7 mm³. The decoupling elements are grounding strips with different lengths. They are placed between the antenna elements in order to achieve wideband decoupling.

A small metal ground is introduced between the two decoupling elements, Strip 1 and Strip 2. This creates a metal boundary between the two decoupling elements. Figure 6.18 shows the E-field distribution between the antenna elements when the metal boundary is not presented. As shown, when either antenna 1 or antenna 2 is excited, there is a strong energy coupled between Strips 1 and 2, even though Strips 1 and 2 have different resonance frequencies. In the example, Strips 1 and 2 were designed to operate at

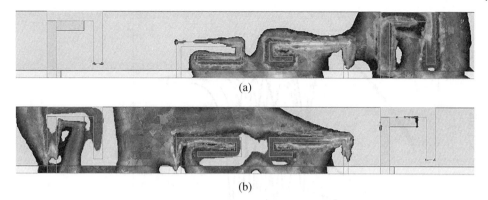

Figure 6.18 Simulated E-field distribution between the antenna elements when the metal boundary is not presented: (a) Antenna 1 is active and (b) Antenna 2 is active. The inactive antenna element is terminated with a matched load.

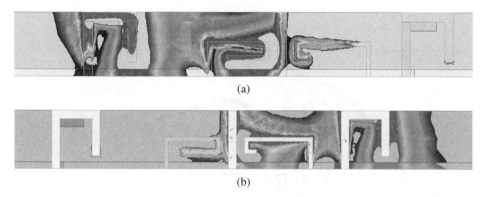

Figure 6.19 Simulated E-field distribution between the antenna elements with the metal boundary: (a) Antenna 1 is active and (b) Antenna 2 is active. The inactive antenna element is terminated with a matched load.

3.46 and 3.58 GHz, respectively. The strong coupling between the decoupling elements means that the EM energy fluctuation of one strip can transmit to the other strip, which leads to disturbance of the EM boundary conditions. Thus, it would be difficult to tune the decoupling elements.

Figure 6.19 shows the E-field distribution between the antenna elements when the metal boundary is placed between the two closely spaced decoupling elements. It can be seen that at the resonant frequency of Strip 1 there is weak or null energy coupled from Strip 1 to Strip 2 when Antenna 1 is active. This means that the small metal ground has blocked the EM energy between Strips 1 and 2. At the resonant frequency of Strip 2, when Antenna 2 is active, the phenomenon is similar. Hence, the EM energy fluctuation of one strip cannot transmit to the other strip so the EM boundary conditions of each strip can be kept stable. Additionally, from the E-field distributions it can be concluded that the decoupling elements operate at $\lambda/4$ mode as monopole antennas.

Figure 6.20 shows the parametric study when the total lengths of the decoupling strips are adjusted. The corresponding parameters are given in Figure 6.17. It can be seen that

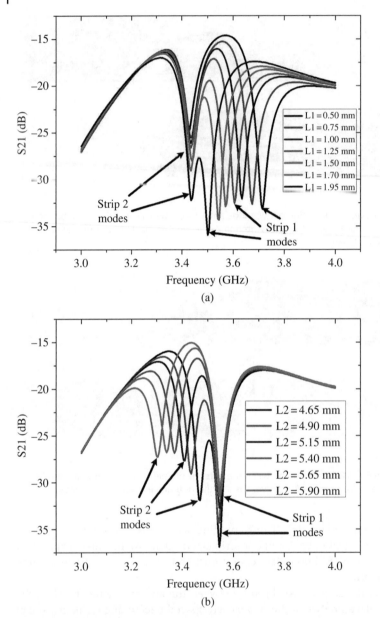

Figure 6.20 Parametric study when the total lengths of the decoupling strips are adjusted: (a) changing the length of decoupling element 1 (Strip 1) and (b) changing the length of decoupling element 2 (Strip 2) [35]. The corresponding parameters are given in Figure 6.17.

the resonant frequencies of Strips 1 and 2 can be independently tuned, which implies that the introduced metal boundary can effectively reduce the mutual coupling between the decoupling elements while maintaining the resonances of the decoupling element without increasing the volume of the MIMO antenna.

6.4 Compact MIMO Antenna with Adaptive Radiation Patterns

In this section, compact MIMO antennas with adaptive radiation patterns are presented. Compared to passive designs, MIMO antennas with adaptive, or reconfigurable radiation patterns can improve the system performance in a more complex environment where the signals show high spatial correlation.

A low-cost adaptive MIMO antenna has been designed based on the folded monopole electronically steerable parasitic array radiator (FM-ESPAR) presented in Chapter 3. The FM-ESPAR antenna presented in Chapter 3 is composed of six folded monopoles and one top-disk-loaded short monopole. The top-disk-loaded monopole is placed at the centre and connected to the RF front end. Six folded monopoles are separated equally in a circle surrounding the driven monopole. In order to achieve a larger capacitive load, the folded monopoles are bent towards the centre. The FM-ESPAR based MIMO antenna is composed of two FM-ESPAR antennas, which are placed on the top and bottom surfaces of a conductive cylinder, as shown in Figure 6.21. In this figure, the two FM-ESPAR antennas in the MIMO antenna are marked as A and B.

The height of the FM-ESPAR antenna is 12.5 mm, which is less than at $\lambda_{2.4GHz}/10$. The cylinder is 30 mm long, which is a quarter-wavelength at 2.5 GHz. Each folded monopole antenna in the FM-ESPAR antenna is connected to the ground plane through a varactor. By controlling the DC voltages applied to the varactors, the radiation pattern of the FM-ESPAR antenna can be steered. The top-disk-loaded monopole at the centre is the driven element. The six folded monopoles surrounding the driven element are parasitic elements. The radiation pattern synthesis is determined by the induced surface currents on all radiating elements. The induced surface currents are tuned by controlling the voltages supplied to the varactors. When supplied with different control voltages, the varactors provide various insertion resistances and capacitances to the

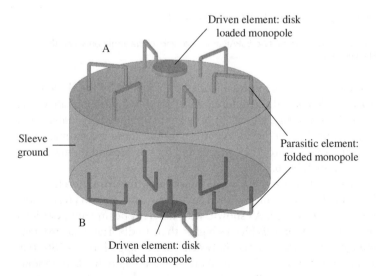

Figure 6.21 The configuration of the ESPAR MIMO antenna presented in [36].

Table 6.2 Beam direction and control voltages (in V) for the FM-ESPAR MIMO antennas [36].

Beam direction	V_1	V_2	V_3	V_4	V_5	V_6
60°	20	20	2	2	2	2
120°	2	20	20	2	2	2
180°	2	2	20	20	2	2
240°	2	2	2	20	20	2
300°	2	2	2	2	20	20
360°	20	2	2	2	2	20

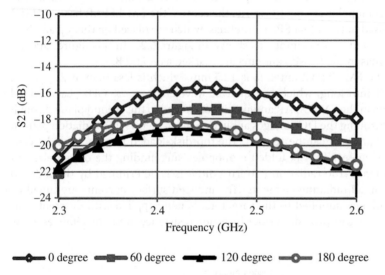

Figure 6.22 Simulated isolation between the two ESPAR antennas versus the angle between their main lobes. Courtesy of Dr Haitao Liu [36].

parasitic elements, therefore various capacitive loads are observed by the driven element. The various capacitive loads and the different insertion resistances can affect the input impedance match of the driven element. In order to maintain a reasonable gain, the FM-ESPAR antenna is matched to the main pattern mode. There are six main patterns. The set up of the varactor's control the voltages and the corresponding maximum radiation direction is shown in Table 6.2.

The FM-ESPAR antennas A and B are able to steer their beams towards different directions individually. The simulated isolations between Antenna A and Antenna B at different scan angles are given in Figure 6.22. As shown, with different beam configurations the isolation between the two antennas is always higher than 15 dB. The measurement results of the radiation pattern when the angle between the two main lobes is 180° (i.e. the main lobes are pointing in opposite directions) are given in Figure 6.23. The radiation patterns of the two antennas are similar to the conventional microstrip patch, with a broad beamwidth and low backlobe.

Figure 6.23 Measured radiation pattern of the MIMO ESPAR antenna in the horizontal plane. Courtesy of Dr Haitao Liu [36].

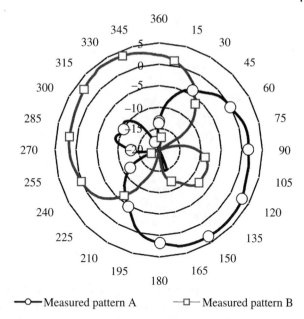

—o—Measured pattern A   —□—Measured pattern B

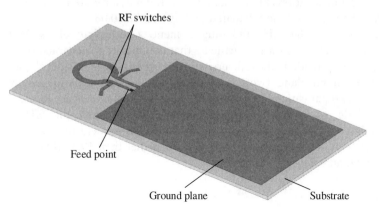

Figure 6.24 Configuration of the radiation pattern reconfigurable monopole [37].

Another approach to realise an adaptive MIMO array is to employ a radiating element that has reconfigurable radiation patterns. For example, by introducing two RF switches to a printed monopole, the radiation patterns of the antenna element can be steered to three different angles [37]. The configuration of the reconfigurable monopole is shown in Figure 6.24. The monopole consists of a circular loop and a bent dipole. The RF switches are located between the loop and the dipole, as shown in Figure 6.24. With the RF switches of different states, the beams of the antenna are steered to 330°, 0° and 40° in the azimuth plane, covering an HPBW angular range of 160°. Within the impedance bandwidth (5.15 to 5.35 GHz) of the antenna, the radiation efficiency of the antenna is higher than 70%.

High isolation between the antenna elements is obtained by placing the monopole at different locations of the PCB board with orthogonal polarisation, as shown in

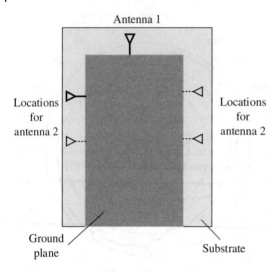

Figure 6.25. The measurement results given in [37] show that for all of these four configurations, in each beam-switching state, this two-element MIMO array has return loss higher than 8.5 dB and isolations higher than 21 dB. The measured total efficiency of the array is higher than 70% and the calculated ECC is less than 0.03.

Besides using a beam-reconfigurable radiating element, a reconfigurable MIMO antenna that utilises two tunable parasitic elements that manipulate the characteristic modes of the structure to provide pattern diversity is reported in [38]. This developed antenna consists of a ground plane, two capacitive exciters located at its diagonally opposite corners, and two parasitic plates where the reconfigurable loads are located. By using the PIN diodes to control the reconfigurable load networks, the MIMO array can have two different sets of radiation patterns. The measurement results show that the isolation between the two ports in both states is higher than 19.5 dB and the calculated ECC is less than 0.5.

6.5 Case Studies

In this case study some techniques to improve the isolation between two closely spaced PIFAs are presented. PIFAs are commonly used in mobile phones because of their compact size, easy integration, and good specific absorption rate properties. Because the available volume to place the antennas is small compared to the wavelength of the antenna, it is always a challenging task to place MIMO antennas on mobile devices and maintain high isolation between each antenna element.

Figure 6.26a shows the typical layout of the PCB for a mobile device. The substrate is FR4 ($\varepsilon_r = 4.4$) with dimension of $150 \times 75 \times 0.8$ mm³. The antennas need to be placed in an area of 6×75 mm. To simplify the analysis, the antenna element is chosen to have a planar structure and the resonance is chosen to be 3.5 GHz. Figure 6.26b shows the configuration of the PIFA and the dimensions of the radiating element. The width of the microstrip is 0.9 mm.

Figure 6.25 Different configurations of the MIMO array presented in [37].

Figure 6.26 (a) Configuration of a single PIFA. (b) Dimensions of a PIFA.

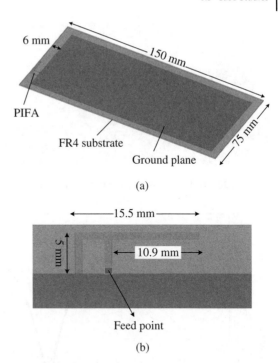

(a)

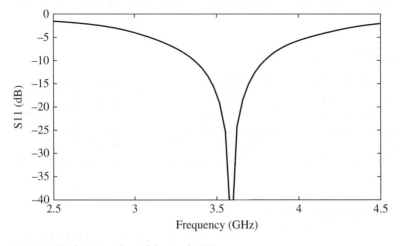

(b)

Figure 6.27 shows the return loss of the PIFA when there is only one PIFA present. The PIFA can be tuned to operate at the desired frequency by adjusting its physical length and the impedance matching can be optimised by changing the offset distance of the feed point. Although good radiation performance is obtained for the single antenna, when another PIFA with the same dimension is introduced, as shown in Figure 6.28, there is high mutual coupling between the two PIFAs, which results in poor isolation between the two antenna ports. The main reason for this high mutual coupling is that the distance between the radiating elements is only 13.5 mm ($0.16\lambda_{3.5GHz}$). Figure 6.29 shows

Figure 6.27 The return loss of the single PIFA.

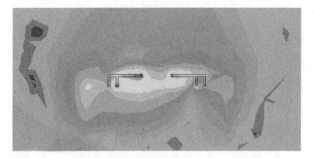

Figure 6.28 The E-field distribution between two PIFAs when one of them is excited.

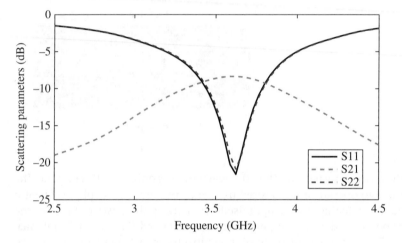

Figure 6.29 The return loss and isolation of two closely spaced PIFAs.

the return loss and isolation of these two PIFAs. As shown, the isolation between the two antenna ports is around 8 dB at the resonance of the antenna. Thus, it is important to improve the isolation between the antenna elements otherwise the MIMO system has a low efficiency.

6.5.1 Increase the Physical Separation

One approach to reducing the mutual coupling between the antenna elements is to increase the physical separation of the antennas. For example, the second PIFA can be placed at the other end of the PCB, as shown in Figure 6.30. Because the distance between the two antennas is relatively large compared to its wavelength, good isolation is obtained, as shown in Figure 6.31. In this case, the distance is more than $1.5\lambda_{3.5GHz}$. The drawback of this method is that it requires additional space to allocate the antenna elements.

6.5.2 Change the Antenna Orientation

Through changing the orientation of the antenna, the polarisation of the PIFA becomes orthogonal, which can lead to increased isolation between the two antennas. As an

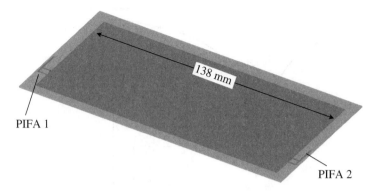

Figure 6.30 The configuration of placing two PIFAs at the two ends of the PCB.

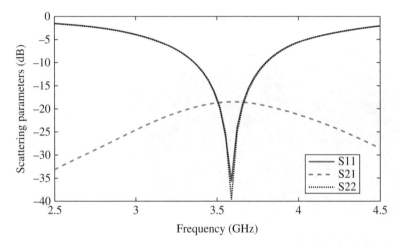

Figure 6.31 The scattering parameters between two antennas when they are placed at different sides of the PCB.

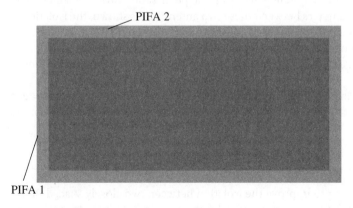

Figure 6.32 Layout of two orthogonally placed PIFAs.

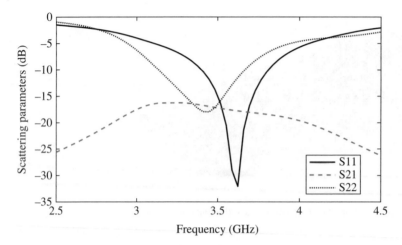

Figure 6.33 The scattering parameters between two antennas when they are placed orthogonally.

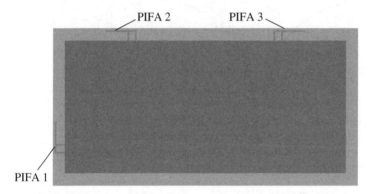

Figure 6.34 Layout for placing three PIFAs.

example, Figure 6.32 shows the layout of two PIFAs that are placed orthogonally. With this configuration, the MIMO system gains from the polarisation diversity. Figure 6.33 shows the scattering parameters between these two antennas. As shown, the isolation is increased to more than 15 dB within the bandwidth of the antenna. It can also be seen that there is some frequency shift from the second PIFA. This is because of the effect of the ground plane on the antenna. The resonance can be re-tuned by adjusting the length of the PIFA.

Because the PCB is relatively long compared to its width, it is also possible to introduce one more PIFA, as illustrated in Figure 6.34. Figure 6.35 shows the isolation between the three PIFAs.

6.5.3 Modify the Ground Plane

Another effective approach to improve the isolation between two closely spaced PIFAs while maintaining good radiation performance for the antenna is to introduce resonant slots on the ground plane [39]. The operation principle is that the resonance structure

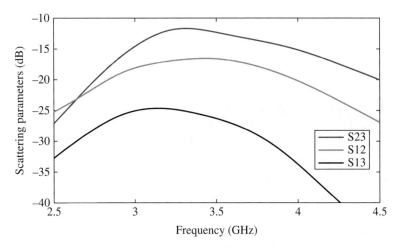

Figure 6.35 The isolation between the three PIFAs shown in Figure 6.34.

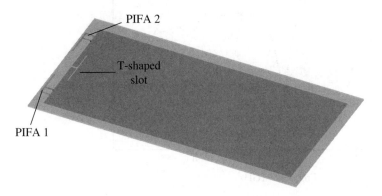

Figure 6.36 Introducing a T-shaped slot between two closely spaced PIFAs.

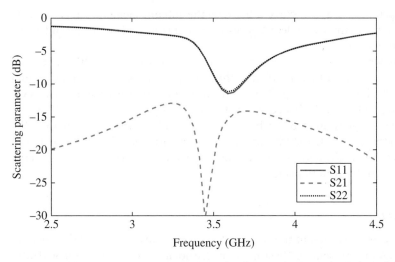

Figure 6.37 The scattering parameters of two closely spaced PIFAs with a T-shaped slot on the ground plane.

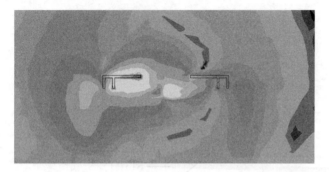

Figure 6.38 The E-field distribution between two PIFAs after introducing the T-shaped slot on the ground plane.

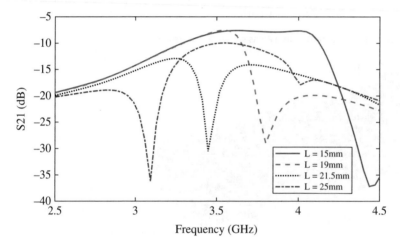

Figure 6.39 The simulated S21 between the two PIFAs when the length of the T-slot varies.

on the ground plane can trap the current between the antenna elements, thus reducing the mutual coupling. In this example, a T-shaped slot is introduced on the ground plane, as shown in Figure 6.36.

Figure 6.37 shows the scattering parameters of the two closely spaced PIFAs. The isolation is improved to higher than 15 dB after the T-shaped slot is etched on the ground plane. Figure 6.38 shows the E-field distribution when one PIFA is excited. As shown, the current is trapped in the T-shaped slot instead of propagating to another PIFA.

Figure 6.39 shows the simulated S21 between the two PIFAs when the length of the T-slot varies. In this parametric study, the length of the horizontal slot is changed from 15 to 25 mm. As shown, the high isolation frequency band is determined by the overall length of the slot on the ground plane.

6.5.4 Summary

In this case study several different simple but effective methods to improve the isolation between two PIFAs are presented. These methods are not limited to the design of MIMO antennas for mobile devices but can be extended to the design of MIMO array antennas for other applications, as such as USB dongles, laptops, tablets, etc.

6.6 Summary of the Chapter

This chapter presents the fundamental theories of MIMO antennas and the related key parameters that can be used to evaluate the performance of the MIMO system. Different techniques for designing low-cost compact MIMO antennas are discussed and design examples are presented. These design techniques can be applied to the practical designs based on the design requirements, including the available volume for the antennas and the required minimum isolation. The case study presented is a good example of how to apply different methods in a practical design as well as having a design trade-off.

References

1 S.L. Loyka. Channel capacity of MIMO architecture using the exponential correlation matrix. *IEEE Communications Letters*, 5(9):369–371, Sept 2001.
2 IEEE. IEEE Standard for Information technology – Local and metropolitan area networks. Specific requirements Part 11: Wireless LAN Medium Access Control (MAC) and Physical Layer (PHY) Specifications Amendment 5: Enhancements for Higher Throughput. *IEEE Std 802.11n-2009 (Amendment to IEEE Std 802.11-2007 as amended by IEEE Std 802.11k-2008, IEEE Std 802.11r-2008, IEEE Std 802.11y-2008, and IEEE Std 802.11w-2009)*, pages 1–565, Oct 2009.
3 M. Shafi, A.F. Molisch, P.J. Smith, T. Haustein, P. Zhu, P. De Silva, F. Tufvesson, A. Benjebbour, and G. Wunder. 5G: A tutorial overview of standards, trials, challenges, deployment, and practice. *IEEE Journal on Selected Areas in Communications*, 35(6):1201–1221, June 2017.
4 L. Zhao and K.L. Wu. A dual-band coupled resonator decoupling network for two coupled antennas. *IEEE Transactions on Antennas and Propagation*, 63(7):2843–2850, July 2015.
5 L. Zhao, L.K. Yeung, and K.L. Wu. A coupled resonator decoupling network for two-element compact antenna arrays in mobile terminals. *IEEE Transactions on Antennas and Propagation*, 62(5):2767–2776, May 2014.
6 K.C. Lin, C.H. Wu, C.H. Lai, and T.G. Ma. Novel dual-band decoupling network for two-element closely spaced array using synthesized microstrip lines. *IEEE Transactions on Antennas and Propagation*, 60(11):5118–5128, Nov 2012.
7 A. Cihangir, F. Sonnerat, F. Ferrero, R. Pilard, F. Gianesello, D. Gloria, P. Brachat, G. Jacquemod, and C. Luxey. Neutralisation technique applied to two coupling element antennas to cover low LTE and GSM communication standards. *Electronics Letters*, 49(13):781–782, June 2013.
8 Q. Luo, C. Quigley, J.R. Pereira, and H.M. Salgado. Inverted-L antennas array in a wireless USB dongle for MIMO application. In *2012 6th European Conference on Antennas and Propagation (EUCAP)*, pages 1909–1912, March 2012.
9 Q. Luo, J.R. Pereira, and H.M. Salgado. Reconfigurable dual-band C-shaped monopole antenna array with high isolation. *Electronics Letters*, 46(13):888–889, June 2010.

10 R. Anitha, V.P. Sarin, P. Mohanan, and K. Vasudevan. Enhanced isolation with defected ground structure in MIMO antenna. *Electronics Letters*, 50(24):1784–1786, 2014.

11 S. Blanch, J. Romeu, and I. Corbella. Exact representation of antenna system diversity performance from input parameter description. *Electronics Letters*, 39(9):705–707, May 2003.

12 P. Hallbjorner. The significance of radiation efficiencies when using S-parameters to calculate the received signal correlation from two antennas. *IEEE Antennas and Wireless Propagation Letters*, 4:97–99, 2005.

13 M. Manteghi and Y. Rahmat-Samii. Multiport characteristics of a wide-band cavity backed annular patch antenna for multipolarization operations. *IEEE Transactions on Antennas and Propagation*, 53(1):466–474, Jan 2005.

14 A. Diallo, C. Luxey, P. Le Thuc, R. Staraj, and G. Kossiavas. Study and reduction of the mutual coupling between two mobile phone PIFAs operating in the DCS1800 and UMTS bands. *IEEE Transactions on Antennas and Propagation*, 54(11):3063–3074, Nov 2006.

15 C. Luxey. Design of multi-antenna systems for UMTS mobile phones. In *2009 Loughborough Antennas Propagation Conference*, pages 57–64, Nov 2009.

16 A. Cihangir, F. Ferrero, G. Jacquemod, P. Brachat, and C. Luxey. Neutralized coupling elements for MIMO operation in 4G mobile terminals. *IEEE Antennas and Wireless Propagation Letters*, 13:141–144, 2014.

17 H.L. Peng, R. Tao, W.Y. Yin, and J.F. Mao. A novel compact dual-band antenna array with high isolations realized using the neutralization technique. *IEEE Transactions on Antennas and Propagation*, 61(4):1956–1962, April 2013.

18 Q. Luo, C. Quigley, J.R. Pereira, and H.M. Salgado. Inverted-L antennas array in a wireless USB dongle for MIMO application. In *2012 6th European Conference on Antennas and Propagation (EUCAP)*, pages 1909–1912, March 2012.

19 S.W. Su, C.T. Lee, and F.S. Chang. Printed MIMO-antenna system using neutralization-line technique for wireless USB-dongle applications. *IEEE Transactions on Antennas and Propagation*, 60(2):456–463, Feb 2012.

20 Z.N. Chen. Note on impedance characteristics of L-shaped wire monopole antenna. *Microwave and Optical Technology Letters*, 26(1):22–23, 2000.

21 Q. Luo, J. Pereira, and H. Salgado. Low cost compact multiband printed monopole antennas and arrays for wireless communications. In Laure Huitema, editor, *Progress in Compact Antennas*, chapter 3. InTech, Rijeka, 2014.

22 S. Zhang and G.F. Pedersen. Mutual coupling reduction for UWB MIMO antennas with a wideband neutralization line. *IEEE Antennas and Wireless Propagation Letters*, 15:166–169, 2016.

23 M.M. Islam, M.T. Islam, M. Samsuzzaman, and M.R.I. Faruque. Compact metamaterial antenna for UWB applications. *Electronics Letters*, 51(16):1222–1224, 2015.

24 K. Li, C. Zhu, L. Li, Y.M. Cai, and C.H. Liang. Design of electrically small metamaterial antenna with ELC and EBG loading. *IEEE Antennas and Wireless Propagation Letters*, 12:678–681, 2013.

25 S. Ahdi Rezaeieh, M.A. Antoniades, and A.M. Abbosh. Gain enhancement of wideband metamaterial-loaded loop antenna with tightly coupled arc-shaped directors. *IEEE Transactions on Antennas and Propagation*, 65(4):2090–2095, April 2017.

26 S. Rezaeieh, M.A. Antoniades, and A.M. Abbosh. Bandwidth and directivity enhancement of loop antenna by nonperiodic distribution of Mu-negative metamaterial unit cells. *IEEE Transactions on Antennas and Propagation*, 64(8):3319–3329, Aug 2016.

27 S.A. Amanatiadis, T.D. Karamanos, and N.V. Kantartzis. Radiation efficiency enhancement of graphene THz antennas utilizing metamaterial substrates. *IEEE Antennas and Wireless Propagation Letters*, 16:2054–2057, 2017.

28 S.H. Hwang, T.S. Yang, J.H. Byun, and A.S. Kim. Complementary pattern method to reduce mutual coupling in metamaterial antennas. *IET Microwaves, Antennas Propagation*, 4(9):1397–1405, September 2010.

29 M. Farahani, J. Pourahmadazar, M. Akbari, M. Nedil, A.R. Sebak, and T.A. Denidni. Mutual coupling reduction in millimeter-wave MIMO antenna array using a metamaterial polarization-rotator wall. *IEEE Antennas and Wireless Propagation Letters*, 16:2324–2327, 2017.

30 D.A. Ketzaki and T.V. Yioultsis. Metamaterial-based design of planar compact MIMO monopoles. *IEEE Transactions on Antennas and Propagation*, 61(5):2758–2766, May 2013.

31 G. Zhai, Z.N. Chen, and X. Qing. Enhanced isolation of a closely spaced four-element MIMO antenna system using metamaterial mushroom. *IEEE Transactions on Antennas and Propagation*, 63(8):3362–3370, Aug 2015.

32 C.H. Wu, C.L. Chiu, and T.G. Ma. Very compact fully lumped decoupling network for a coupled two-element array. *IEEE Antennas and Wireless Propagation Letters*, 15:158–161, 2016.

33 X.Y. Zhang, C.D. Xue, Y. Cao, and C.F. Ding. Compact MIMO antenna with embedded decoupling network. In *2017 IEEE International Conference on Computational Electromagnetics (ICCEM)*, pages 64–66,March 2017.

34 S. Zhang, Z. Ying, J. Xiong, and S. He. Ultrawideband MIMO/diversity antennas with a tree-like structure to enhance wideband isolation. *IEEE Antennas and Wireless Propagation Letters*, 8:1279–1282, 2009.

35 H. Xu, H. Zhou, S. Gao, H. Wang, and Y. Cheng. Multimode decoupling technique with independent tuning characteristic for mobile terminals. *IEEE Transactions on Antennas and Propagation*, 65(12):6739–6751, Dec 2017.

36 H. Liu. *Design and Measurement Methodologies of Low-Cost Smart Antennas*. PhD thesis, Faculty of Engineering and Physical Sciences, University of Surrey, May 2012.

37 C. Rhee, Y. Kim, T. Park, S.S. Kwoun, B. Mun, B. Lee, and C. Jung. Pattern-reconfigurable MIMO antenna for high isolation and low correlation. *IEEE Antennas and Wireless Propagation Letters*, 13:1373–1376, 2014.

38 K.K. Kishor and S.V. Hum. A pattern reconfigurable chassis-mode MIMO antenna. *IEEE Transactions on Antennas and Propagation*, 62(6):3290–3298, June 2014.

39 J. Park, J. Choi, J.Y. Park, and Y.S. Kim. Study of a T-shaped slot with a capacitor for high isolation between MIMO antennas. *IEEE Antennas and Wireless Propagation Letters*, 11:1541–1544, 2012.

7

Other Types of Low-cost Smart Antennas

7.1 Introduction

In this chapter, some other types of low-cost smart antennas are presented. These include the multi-beam dielectric lens, retrodirective array, pattern-reconfigurable FP resonant antenna, beam-switching antennas using the AFR, and multi-beam antennas based on passive beamforming networks. The dielectric lens, FP resonant antenna, and AFR provide low-cost solutions to achieving high directivity and multiple beams (or beam-switching) without resorting to the use of a large number of active radiating elements, thus they are attractive solutions to microwave and mm-wave smart antennas, such as smart antennas for 5G massive MIMO base stations, mobile satellite communications, and automotive radars. Multibeam or beam-switching antennas can also be achieved by using antenna arrays integrated with low-cost passive beamforming networks such as the Butler matrix, Nolen matrix, and Rotman lens. The retrodirective array determines the angle of arrival of the incident signal and re-transmit the signal back without using any phase shifters, leading to a significant reduction in the cost of the antenna system. In the following sections, the principles and design techniques of these low-cost smart antennas are explained and examples from recent reported designs are presented.

7.2 Lens Antennas

7.2.1 Lens Antenna Basics

A lens antenna is a microwave antenna in which an antenna or array antenna is placed in front of a dielectric lens. The lens concentrates the radiated waves from the antenna to form a focused beam in the far-field. The lens can collimate a spherical or cylindrical wavefront produced by the feed into an outgoing planar or linear wavefront. Because the high directivity is a result of the different phase velocity in air and in the lens material, the lenses can be categorised as slow wave and fast wave lenses, as shown in Figure 7.1 [1]. The fast-wave lens has shorter effective electrical path while the slow-wave lens increases its effective electrical path length in order to let all rays have an equal phase front after travelling through the lens.

Low-cost Smart Antennas, First Edition. Qi Luo, Steven (Shichang) Gao, Wei Liu, and Chao Gu.

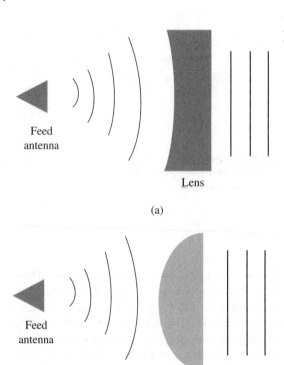

Figure 7.1 Different types of lens antenna: (a) fast wave lens and (b) slow wave lens.

Feed antenna

Lens

(a)

Feed antenna

Lens

(b)

The lens curvature can be calculated by using geometric optics. The ray tracing approach is used to form the equations on the condition that the waves experience equal electrical path length when propagating through the lens. Figure 7.2 shows the configuration of a surface refracting lens. To meet the condition of equal electrical length, the following equations are derived [1]

$$nL + R = nT + F \tag{7.1}$$

$$R = \frac{(n-1)F}{n\cos\theta - 1} \tag{7.2}$$

where L and R are the arbitrary paths in the air and in the dielectric, respectively, F is the focal distance of the lens and T is the maximum lens thickness. n is the refractive index, which is

$$n = \sqrt{\varepsilon_r \mu_r} \tag{7.3}$$

where ε_r is the material's relative permittivity and μ_r is the relative permeability. The maximum lens thickness T is related to the diameter of the lens and is given by

$$\frac{T}{D} = \sqrt{\frac{1}{n^2}\left(\frac{F}{D}\right)^2 + \frac{1}{4n(n+1)}} - \frac{F}{nD} \tag{7.4}$$

Figure 7.2 The configuration of a surface refracting lens.

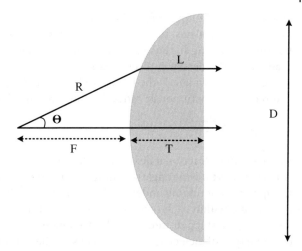

Figure 7.3 The configuration of another type of surface refracting lens.

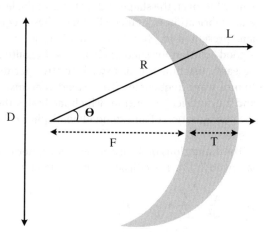

Another type of lens that has only one refracting surface is shown in Figure 7.3. Different from the lens shown in Figure 7.2, the surface facing the feed is spherical and the refraction takes place on the opposite side of the feed, whose profile can be described by the polar equation [1]

$$r = \frac{(n-1)F}{\cos\theta + n} \tag{7.5}$$

where the maximum lens thickness T is

$$\frac{T}{D} = \frac{1}{2\sqrt{(n^2 - 1)}} \tag{7.6}$$

7.2.2 Millimetre-wave Lens Antenna Design

Using a lens is an effective approach to obtaining a high directivity antenna design with low sidelobes. Meanwhile, beam-switching or a multi-beam can be achieved by placing multiple feed antennas in front of the dielectric lens. Due to its high mass and large size, lens antennas are not preferred in the design of low-frequency antennas. However, at mm-wave frequencies, the volume and weight of the lens become acceptable because of the small wavelength, thus mm-wave lens antennas have attracted much interest from academia and industry in recent years.

The best-known lens is the Luneberg lens. It has a radially symmetrical configuration and requires a continuously refractive index profile given by the relation [2]

$$n = \sqrt{2 - r^2} \tag{7.7}$$

where n is the refraction index and r is the radius of the lens. Although the Luneberg lens is an ideal lens for wide angle scanning applications, the construction is complex because it requires a continuous index profile. One approach is to divide the lens into several regions and within each region the dielectric of the same permittivity is applied. Thus, the lens has a stepped-index profile. To obtain a low-cost design, it is desirable to use only one type of dielectric material to design the lens instead of having a stepped-index lens. Moreover, the shape and material of the lens should be easily machined with moderate fabrication tolerance. To meet these requirements, the extended hemispherical lens presented in [3], which is made of a low-cost plastic material Rexolite, provides a good solution for such applications. Rexolite is a type of low-cost substrate that can be easily machined. It has stable relative permittivity (ε_r) of 2.54 from low frequency to mm-wave frequencies. The extended hemispherical lens consists of half of an ellipse and a cylindrical extension, which facilitates the machining process. Figure 7.4 shows the configuration of an extended hemispherical lens with a microstrip patch antenna as the feed.

The dimensions of the extended hemispherical lens are calculated by using the formulas given in [3]. The ellipse is represented by

$$\frac{x^2}{a^2} + \frac{y^2}{a^2} + \frac{z^2}{b^2} = 1 \tag{7.8}$$

where a and b are the minor and major axes of the ellipsoidal lens, respectively. The a and b should satisfy

$$a = b \times \sqrt{\frac{\varepsilon_r - 1}{\varepsilon_r}} \tag{7.9}$$

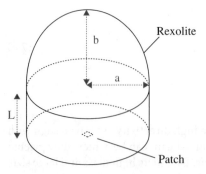

Figure 7.4 Configuration of an extended hemispherical lens using a microstrip patch antenna as the feed.

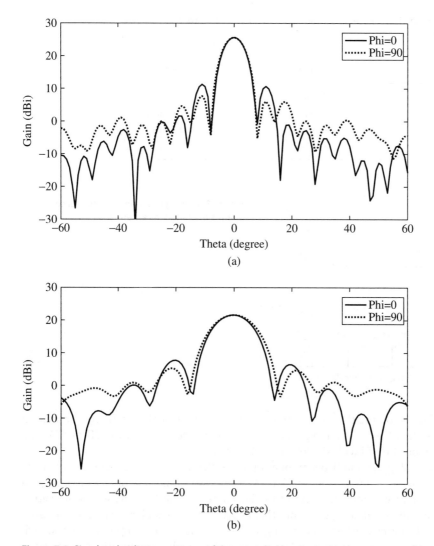

Figure 7.5 Simulated radiation patterns of the extended hemispherical lens antenna shown in Figure 7.4: (a) Design A and (b) Design B.

where ε_r is the relative permittivity of the material. The radius of the cylindrical extension is equal to the minor axes of the ellipse and its height is

$$L = \frac{b}{\sqrt{\varepsilon_r}} \tag{7.10}$$

To avoid the unwanted radiation from the microstrip feed line and facilitate antenna integration, it is better to use an aperture-coupled patch as the feed antenna. The directivity of the antenna is proportional to the dimension of the lens. Figure 7.5 shows the simulated radiation patterns of the extended hemispherical lens antenna of different dimensions (Designs A and B). The feed antenna is a linearly polarised square patch that resonates at 30 GHz. The parameters of these two lens antennas are given in Table 7.1.

Table 7.1 Comparison of the presented design with other reported wideband transmitarrays.

Parameter	Design A	Design B
Material	Rexolite	Rexolite
a	38.9 mm	19.5 mm
b	50 mm	25 mm
L	31.4 mm	15.7 mm

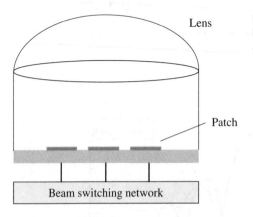

Lens

Patch

Beam switching network

Figure 7.6 Using multiple feeds to realise a beam-switching lens antenna.

Design A has a volume that is four times as large as that of Design B. The simulation results show that Design A has a gain of 25.7 dBi with sidelobe level 15.7dB, and Design B has a gain of 21.6 dBi with sidelobe level 15.2 dB.

Beam-switching or multi-beam operation of the lens can be realised by using multiple feeds, as shown in Figure 7.6. A beam-switching network is incorporated with the feed arrays, and the beam of the lens antenna can be switched by activating the corresponding radiating element. This approach is particularly useful to design a low-cost high-directivity beam-switching smart antenna at mm-wave frequencies, such as the E-band antenna system for 5G access and the backhaul applications reported in [4]. In the design presented in [4], a number of TriQuint/Qorvo TGS4306-FC SP4T switches were used to design the feed network. These SP4T switches form a three-level beam-switching, as shown in Figure 7.7.

Similar to planar arrays, a feed array can have either a rectangular or a hexagonal lattice, as shown in Figure 7.8. Figure 7.9 shows the beam-switching radiation patterns of the Design A lens by placing a linear four-element array as the feed. The distance between the radiating elements is 7.5 mm, which is $0.75\lambda_{30GHz}$. By activating the corresponding element, the beam of the lens is switched from broadside to 27°, with the gain varied from 26.7 to 21.7 dBi, representing a scan loss of 5 dB. The beam is switched at an angular step of about 10° and the beam overlap is around −5 dB. To achieve a higher beam-scanning resolution, the distance between the feed element needs to be reduced. Figure 7.10 shows the simulated scanning radiation patterns of the lens antenna obtained by reducing the distance between the feed element to $0.5\lambda_{30GHz}$. As shown, the

Figure 7.7 Concept of the beam-switching feed network based on PIN diode switches [4].

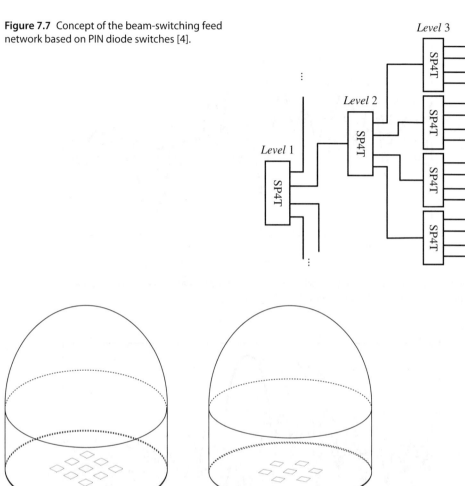

Figure 7.8 Extended hemispherical lens antenna with different feed array: (a) rectangular lattice and (b) hexagonal lattice.

beam of the lens antenna can be switched to 20°, and the gain is varied from 26.2 to 23.4 dBi. After reducing the distance between the feed elements, the beam overlap is increased to −3 dB.

The lens antenna provides a cost-effective solution to mm-wave smart antennas, including MIMO systems [5]. In this section, two basic lens antenna designs are presented. Lens antenna design is a comprehensive topic and there are many different types of lens, some of them more complicated than the two examples shown in this section. Since the purpose of this section is to introduce the use of the lens to design low-cost smart antennas at mm-wave frequencies, it is impossible to cover all types of lens antenna in this section and the interested reader can refer to [1] for more advanced topics in lens antenna design.

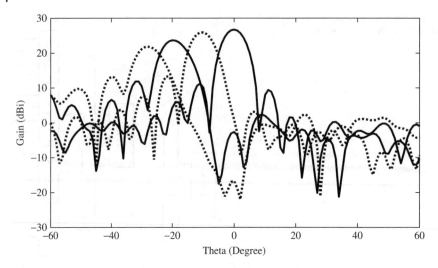

Figure 7.9 The beam-switching radiation patterns of the lens antenna fed by a linear array. The spacing between the feed elements is $0.75\lambda_{30GHz}$.

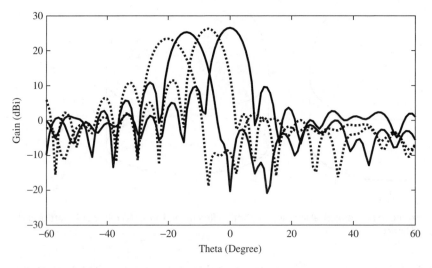

Figure 7.10 The beam-switching radiation patterns of the lens antenna fed by a linear array. The spacing between the feed elements is $0.5\lambda_{30GHz}$.

7.3 Retrodirective Array Antenna

The retrodirective array is an array antenna that can receive the signal from the interrogator and re-transmit a signal back to the interrogator without any prior knowledge of the incoming direction of the receiving signal [6]. Unlike traditional phased arrays, which rely on sophisticated signal-processing algorithms and beamforming networks to form the shaped beam to the desired direction, the retrodirective array performs self-beam tracking with a much simpler system. Thus, the retrodirective array is an attractive solution to the design of low-cost smart antennas.

General speaking, there are two types of retrodirective array. The first type is the Van Altta array [7] and the other type is based on the heterodyne method [8]. Applications of retrodirective arrays include satellite communications, road traffic management, wireless power transmission, etc.

7.3.1 Van Altta Array

The Van Altta array consists of an array of radiating elements interconnected in pairs, as shown in Figure 7.11. The interconnected antenna elements are equidistant from the array centre and each of the transmission lines that connects the two antenna elements must have an equal phase delay. When the Van Altta array receives the incident wave, each of the radiators transmits its received signal through the interconnecting transmission line to feed another radiator that is symmetrical to the centre of the array and re-radiates the waves. The re-radiated wave has a reverse phase distribution compared to the incident wave (see Figure 7.11), which results in a shaped beam pointing at the angle of the incident wave.

Because the Van Atta array is more suitable for planar topologies [9], the form of the radiator is normally chosen as either a microstrip patch or a slot antenna. In [10], a Van Atta retrodirective reflector was designed and developed using aperture-coupled microstrip antennas. The array consists of three identical subarrays, each of which has four antenna elements. The antenna pairs were connected by equal length microstrip lines. The configuration of this array antenna is shown in Figure 7.12. As shown, because

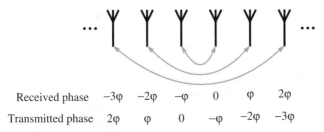

Received phase	-3φ	-2φ	$-\varphi$	0	φ	2φ
Transmitted phase	2φ	φ	0	$-\varphi$	-2φ	-3φ

Figure 7.11 The concept of the Van Altta linear array antenna.

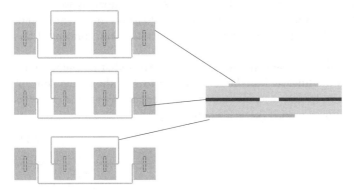

Figure 7.12 The configuration of the Van Altta array using aperture-coupled microstrip patches [10].

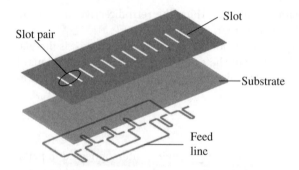

Figure 7.13 The configuration of a retrodirective array using dual-slot antenna elements [11].

the patch and the microstrip feed lines are printed on different layers, there is more space to route the microstrips, which makes this type of configuration a preferred approach to designing a microstrip Van Atta array.

A planar retrodirective array using dual-slot antennas was reported in [11]. The radiator is a dual-slot antenna for the purpose of increasing the overall antenna gain. The radiators are paired symmetrically, as shown in Figure 7.13. One advantage of using a slot antenna as the radiating element is that the microstrip lines which are used to interconnect the antenna element are printed behind the ground plane, so the spurious radiation from the microstrip lines can be suppressed. Since the slot antenna can also radiate backwards, a ground plane is usually required to work as the reflector.

At millimetre frequencies, the loss of the microstrip lines becomes high. Moreover, there is limited available space to route the microstrip feeds line and the couplings between the feed lines are increased due to the small distance between the radiating elements. The coupling between the feed lines should be minimised as it affects the impedance of the microstrip line and introduces unwanted phase delays, which affects in the beam-pointing accuracy of the re-transmitted waves. To address this issue, a substrate integrated waveguide (SIW) was used to design the Van Atta array. In [12], a 30 GHz SIW passive Van Atta array was presented. This SIW array uses four alternating longitudinal slots as the radiating elements and each of the radiating elements is fed by equal length SIW feeding lines. A retrodirective antenna array that can rotate the polarisation was presented in [13]. This antenna was also realised using SIW technology. In this study, two different subarrays were used to receive and transmit the signal. The slot antenna is used as the radiating element and the radiators at one subarray were interconnected to the radiators of another subarray. In the two subarrays the orientation of the slots is orthogonal, which leads to the polarisation conversion. The concept of this design is shown in Figure 7.14.

The Van Altta array can also be applied to the planar array. The radiating elements are interconnected in pairs as the linearly Van Altta array. Figure 7.15 shows the concept of the planar Van Altta array. In this case, for example, the radiator located at (x_1, y_1) is connected to the radiator positioned at (x_2, y_2), where

$$x_1 = -x_2 \tag{7.11}$$

$$y_1 = -y_2 \tag{7.12}$$

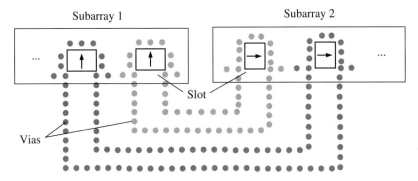

Figure 7.14 The configuration of the polarisation convert SIW Van Altta array [12].

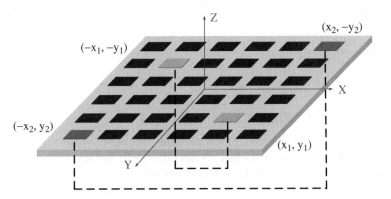

Figure 7.15 The configuration of the planar Van Altta array antenna.

7.3.2 Phase Conjugating Array

Another type of retrodirective array is realised using the concept of phase conjugation. In this approach, instead of interconnecting two radiators to achieve the reversal phase gradient, additional circuitry is added to each antenna element to reverse the sign of the phase, so the re-transmit signal has a phase which is conjugated to the phase of the received signal. A popular method for obtaining the conjugated phase is to use the heterodyne technique [8].

Figure 7.16 shows a typical configuration of retrodirective arrays using the heterodyne technique. As shown, each of the antenna elements is connected to a mixer, where the received RF signal is mixed with a local oscillator (LO). The mixed product is [9]

$$
\begin{aligned}
V_{IF} &= V_{RF}\cos(\omega_{RF}t + \varphi)V_{LO}\cos(\omega_{LO}t) \\
&= \tfrac{1}{2}V_{RF}V_{LO}[\cos((\omega_{LO} - \omega_{RF})t - \varphi) + \cos((\omega_{LO} + \omega_{RF})t + \varphi)]
\end{aligned}
\tag{7.13}
$$

where φ is the phase of the received RF signal, and ω_{RF} and ω_{LO} are the frequencies of the received signal and the LO, respectively. From the equation, it can be seen that if the LO frequency is twice of the frequency of the received RF signal ($\omega_{LO} = 2\omega_{RF}$), then

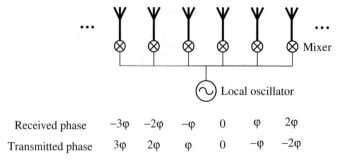

Figure 7.16 The configuration of the retrodirective array using the heterodyne technique.

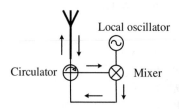

Figure 7.17 Self-transmitting retrodirective antenna element using a circulator and a mixer.

$$V_{IF} = \frac{1}{2} V_{RF} V_{LO} [\cos(\omega_{RF} t - \varphi) + \cos(3\omega_{RF} t + \varphi)] \tag{7.14}$$

In this way, the intermediate frequency (IF) has the same frequency as the received RF signal but with a conjugated phase. The other product has the frequency of $3\omega_{RF}$ and it can be easily filtered out since it is quite far from the desired signal.

Because the IF signal and the received RF signal have the same frequency, to avoid the mutual interference between the leaked RF signal and the transmitted signal a small frequency offset can be introduced so that the RF and IF signals do not have the same frequency [14]. This can be achieved by choosing the frequency of the LO slightly larger or smaller than the $2\omega_{RF}$. However, the disadvantage of this approach is that the frequency difference ($\Delta f = AF - IF$) between the RF and IF signals causes some beam pointing error; thus, the frequency of the LO should be carefully selected. The re-radiated wave has a beam squint, which can be calculated as [8]

$$\theta = \sin^{-1}\left(\frac{f_{in}}{f_{out}} \sin\phi\right) \tag{7.15}$$

where f_{in} is the frequency of the receiving signal, f_{out} is the frequency of the output signal, and ϕ is the angle of the beam. In practical design, if the same radiating element is used to receive and re-transmit the RF signal, a circulator which acts as an isolator between the received and re-transmitted signals is required, as shown in Figure 7.17.

Similar to phased arrays, when there is a large number of antenna elements, the feed network of the retrodirective array, e.g. a corporate-fed transmission line network, becomes very complex. An interesting concept was presented in [15], where the space-fed concept is applied to feed the LO. This has the same principle as used in the reflectarray and transmitarray designs (see Chapter 5). Figure 7.18 presents the concept of the retrodirective array using space-fed local oscillators.

As the circulator is normally bulky and costly, using a large number of circulators increases the cost of the smart antenna system. One approach to avoid the use of

Figure 7.18 The concept of the retrodirective array using space-fed local oscillators [15].

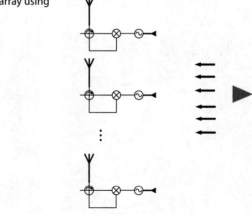

Figure 7.19 The schematic of the active retrodirective antenna presented in [17]. Reproduced with permission of ©EurAPP.

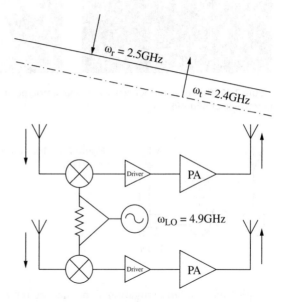

circulators is to use two subarrays to separately receive and transmit the RF signal [16]. This configuration is similar to the Van Altta array, where antenna elements are grouped in pairs. Figure 7.19 shows the configuration of the retrodirective antenna with an active transmitter module [17]. The receiving and transmitting frequencies are offset by 100 MHz to improve the RF-to-IF isolation. Thus, the LO frequency is chosen to be 100 MHz smaller than twice the receiving frequency. As this retrodirective antenna is designed for the wireless power transfer application, power amplifiers are introduced in this system for the purpose of increasing the radiated RF power. In this design, the mixer used is an LTC5549 model from Linear Technology and the LO is a voltage-controlled oscillator (VCO) ZX95-5400+ from Mini-circuits. Photographs of the fabricated prototype are shown in Figure 7.20.

Using the same concept, the retrodirective array can be designed as a dual-band system by letting the receiving and transmitting subarrays operate at different frequencies. A low-frequency interrogating signal can then be retransmitted at a higher frequency or

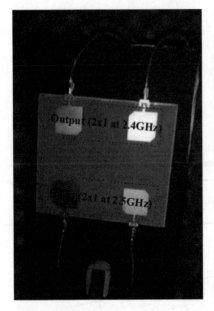
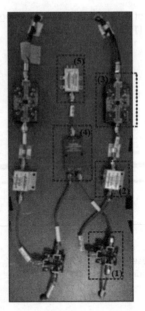

Figure 7.20 The active retrodirective antenna prototype presented in [17]. Reproduced with permission of ©EurAPP.

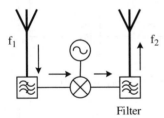

Figure 7.21 The concept of the dual-frequency retrodirective array [18].

a high-frequency interrogating signal can be retransmitted at a lower frequency. If the antennas of different frequencies are positioned orthogonally, polarisation conversion can also be obtained. Figure 7.21 illustrates the concept of the dual-frequency retrodirective array presented in [18]. This array antenna operates at 2.44 and 6 GHz. Filters are used after or before the antennas. The filter can also be integrated into the antenna design, e.g. filter-antenna. In this way, the antenna and filter can be realised in the form of microstrips, which can keep the cost of the system low.

To further reduce the cost of the antenna, a retrodirective transponder array with a time-shared phase conjugator is proposed in [19]. In this approach, by introducing a switching network to the antenna elements, a single-phase conjugator can be sequentially time shared by all of the antenna elements, as shown in Figure 7.22. Because there is only one antenna element to access the phase conjugator at a time, the amount of required phase-conjugating circuitry is reduced by a factor of N, where N is the number of antenna elements in the array. As a result of the shared mixer, the output waveform of each antenna is similar to the amplitude shift keying type of modulated waveform. For

Figure 7.22 The concept of a time-shared phase conjugator retrodirectarray [19]

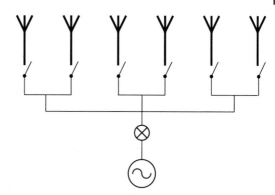

an array antenna of N elements, the waveform of each channel is

$$x_n(t) = \left\{ \begin{array}{l} 1, if\, \frac{(n-1)T_s}{N} < t < \frac{nT_s}{N} \\ 0, else \end{array} \right\} \tag{7.16}$$

where T_s is the time for all antennas to access the phase conjugator and $x_n(t)$ is the amplitude of the output signal. The signal that is re-transmitted is the sum of each channel [19]

$$S_n(t) = x_n(t).\frac{1}{2}V_{RF}.V_{LO}.[\cos(\omega_{IF}t - \theta_{RFn})] \tag{7.17}$$

The switching speed of the RF switches makes sure that the phase conjugation process takes place almost instantaneously, which lets the re-transmitted signal point to the desired direction. However, because each channel only has a limited time to transmit, the power of the re-transmitted signal is small compared to the conventional retrodirective array, which is the trade-off of this approach. To compensate for the power loss, as pointed out by [19], an active mixer or active circulator can be used.

7.4 Fabry–Perot Resonator Antennas

An FP cavity antenna consists of a feed antenna located in a resonant cavity formed between a reflector and a partially reflective surface (PRS). Figure 7.23 shows the configuration of a typical FP cavity antenna.

Ray optics theory can be used to explain the operation principle of the FP cavity antenna. The directivity of the antenna is significantly increased through multiple reflections between the ground plane and the PRS. To reach the resonant condition, the cavity height of the FP cavity antenna needs to satisfy [20]

$$h = \frac{\lambda}{4\pi}(\phi_{PRS} - \pi) + \frac{\lambda}{2} \times N, N = 0, 1, 2, ... \tag{7.18}$$

where h is the cavity height, λ is the wavelength of the operating frequency, and ϕ_{PRS} is the reflection phase of the PRS.

With the development of periodic structures such as the AMC, which can provide approximate zero reflection phase at the frequency of interest, the ground plane of the FP cavity antenna can be replaced by the AMC and the profile of the antenna can be

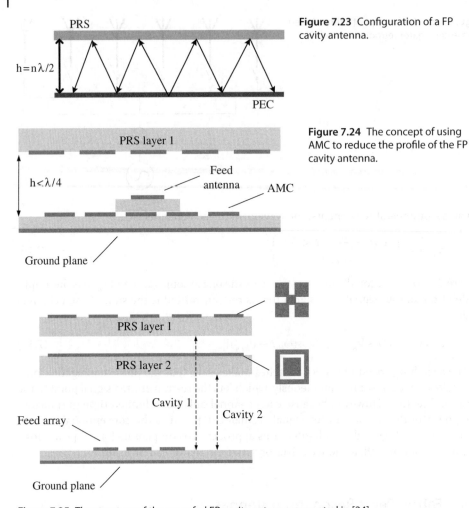

Figure 7.23 Configuration of a FP cavity antenna.

Figure 7.24 The concept of using AMC to reduce the profile of the FP cavity antenna.

Figure 7.25 The structure of the array-fed FP cavity antenna presented in [24].

reduced to less than $\lambda/4$ [21, 22]. Figure 7.24 shows the concept of using AMC to reduce the profile of the conventional FP cavity antenna.

Because the FP cavity antenna achieves high directivity with a small number of active radiating elements compared to the traditional phased array, it has become an attractive candidate for the design of low-cost smart antennas. To steer the beam, one approach is to employ a phased array as the feed antenna and obtain a region of the phase gradient on the aperture of the PRS. Another benefit of using array antenna to feed the FP cavity antenna is that it can improve the bandwidth [23], which is a drawback of the traditional FP antennas. The shared-aperture dual-band FP cavity antenna presented in [24] is a good example of this approach. Figure 7.25 shows the structure of this array-fed FP cavity antenna.

Two frequency selective layers are mounted above the ground plane with different heights to form two separate FP cavities. Each cavity operates at one resonant frequency. The upper layer uses patch-type FSS and the lower layer uses ring-slot type FSS as unit cells. As a thumb of rule, each FSS layer should satisfy the condition that it has high

Figure 7.26 The layouts of the feed array unit cell presented in [24].

reflection magnitude at its operating frequency while it is almost transparent to other frequencies. When this antenna is working at the lower frequency, where the resonant cavity is formed by the upper FSS and the ground plane, the lower FSS is almost transparent to the waves at the lower frequency. When this antenna operates at the upper frequency, the upper FSS has little effect on the propagation of the waves. Figure 7.26 shows the layouts of the feed array unit cell. The C-band patch is placed in the centre and two X-band elements are symmetrically placed at the two sides of the C-band patch. To obtain an acceptable isolation, the spacing between two X-band elements is chosen as 40 mm. It is shown in [24] that by using four array elements as the feed array, which means that the feed array for the X-band is a 1×8 linear array and for the C-band is a 1×4 linear array, the beams of the antenna at either frequency can be steered within $\pm 15°$ with reasonable sidelobe levels (Figure 7.27).

It is noted that there are some disadvantages of this approach, such as limited scan angle and relatively high sidelobe. The scan angle is limited by the phase gradient on the aperture of the PRS, which shows higher phase errors at large scan angles. Meanwhile, the FP cavity also amplifies the sidelobe of the feed array. Thus, it is important to taper the excitation coefficients of the feed array in order to suppress its sidelobes. Another approach to obtaining an electronically beam steerable FP leaky-wave antenna is to employ a tunable HIS [25, 26]. Figure 7.28 shows the configuration of the antenna presented in [25]. This leaky-wave antenna is based on a parallel-plate waveguide loaded with a planar partially reflective surface and a high impedance surface. The HIS is integrated with varactors and by controlling the biasing voltage on the varactors, which changes the capacitance of the HIS, the scattering properties of the HIS can be controlled. As a result, the beam of this antenna can be scanned from broadside towards the endfire direction.

Figure 7.29 shows the simulated reflection phase of the tunable HIS with an incident plane wave of different incidence angles. As shown, with varied value of the capacitance (C_j), the reflection phase of the HIS varies from $150°$ to $-180°$.

Figure 7.30 shows the computed H-plane normalised directivity radiation patterns for the active beam-steering FP leaky-wave antenna at 5.6 GHz. When the value of C_j varies from 0.01 to 0.23 pF, the beam continuously scans from $5°$ to $50°$.

Instead of tuning the reflection phase of the HIS, it is demonstrated in [27] that by adjusting the transmission coefficient phase of the PRS, the beam of the FP antenna can be steered within $\pm 20°$. Although in this study the phase shift was obtained by changing the spacing between each PRS unit cell of the capacitive grid, it has the potential to be developed as a mechanical steering or electronic steering FP antenna.

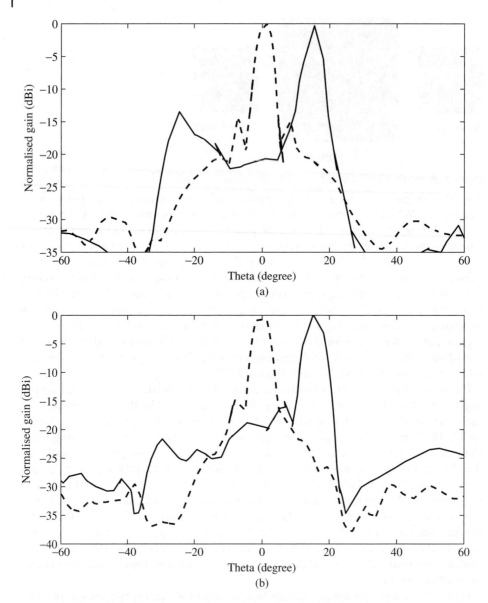

Figure 7.27 The calculated scanning pattern in (a) the C-band and (b) the X-band.

7.5 Array-fed Reflector

7.5.1 Operation Principle

A reflector antenna consists of one or more reflecting surfaces and a feed system for transmitting and/or receiving electromagnetic waves [28]. There are several types of reflector antenna, including central-fed, offset-fed, Cassegrain, and Gregorian. Central-fed and offset-fed reflectors use a single reflector while Cassegrain and

Figure 7.28 The configuration of the reconfigurable FP leaky-wave antenna presented in [25].

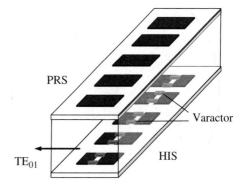

Figure 7.29 Simulated reflection phase of the tunable HIS with an incident plane wave of different incidence angles [25]. Reproduced with permission of ©EurAPP.

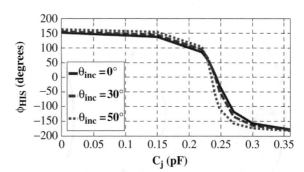

Figure 7.30 The computed H-plane normalised directivity radiation patterns for the active beam-steering FP leaky-wave antenna at 5.6 GHz [25]. Reproduced with permission of ©EurAPP.

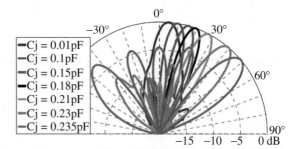

Gregorian reflectors employ a secondary reflector to re-reflect the incident wave from the feed antenna in order to increase the effective focal length of the reflector. Figure 7.31 shows the configurations of these four types of reflector antenna.

The directivity of the reflector is determined by its electrical size and can be calculated by

$$D = \frac{4\pi}{\lambda^2}A \tag{7.19}$$

where λ is the free-space wavelength and A is the physical size of the aperture. The gain of the reflector is

$$G = \varepsilon_{ap}D = \varepsilon_{ap}\frac{4\pi}{\lambda^2}A \tag{7.20}$$

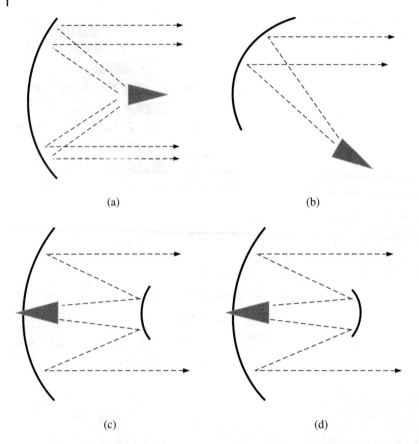

(a)

(b)

(c)

(d)

Figure 7.31 Different types of reflector antenna: (a) central-fed, (b) offset-fed, (c) Cassegrain, and (d) Gregorian.

where ε_{ap} is the aperture efficiency of the reflector. The aperture efficiency of the reflector is related to several factors [29], including:

- taper efficiency ϵ_t: the uniformity of the amplitude distribution of the feed pattern over the surface of the reflector
- spillover efficiency ϵ_s: the fraction of the total incident power that is reflected by the reflector
- phase efficiency ϵ_p: the phase uniformity of the field over the aperture plane
- polarisation efficiency ϵ_x: the polarisation uniformity of the field over the aperture plane
- blockage efficiency ϵ_b: the fraction of the radiated power from the reflector blocked by other objects, such as the mounting structure or the feed antenna
- random error ϵ_r: the random error on the reflector surface, such as the imperfect surface.

The reflector antenna has the advantages of high efficiency and high directivity, thus it is widely used in satellite communications. Compared to phased arrays, the cost of reflector antennas is low. Although it has a high profile and large mass, the reflector

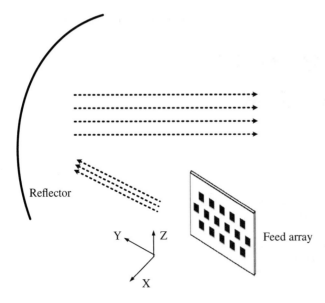

Figure 7.32 The configuration of the AFR with a phased array as the feed.

antenna is still very attractive for satellite and ground station applications. By using a planar array as the feed, the AFR can electronically switch beams or realise multi-beams. This approach allows for fast beam-switching with a relatively simple beamforming network, and the use of a small number of RF chains to achieve high gain. These advantages significantly reduce the cost of the smart antenna system, especially at mm-wave frequencies. Figure 7.32 shows the configuration of an AFR with offset feeding. As shown, an array antenna is placed at the focal point of the reflector to replace the traditional horn antenna. Because the array antenna has larger lateral dimensions, to avoid the blockage effect offset-feed is employed.

Figure 7.33 shows the top view of the feed array and its system architecture. The feed array consists of several subarrays and by switching between these subarrays the beams of the reflector are switched to different angles. In this example, there are 16 radiating elements which can form four feed subarrays. The array elements are aperture-coupled stacked patches [30] positioned in a hexagonal lattice separated by a distance of around 0.8λ.

Figure 7.34 shows the simulated beam-switching radiation patterns of this AFR. The diameter of the reflector is 300 mm and the central frequency of the feed is 20 GHz. By switching between these four feeding clusters, the AFR is able to switch its beam to $6°$ with gain variation less than 0.5 dB. The directivity of the broadside beam is 34 dBi with a 3 dB beam width of $3.6°$. During the beam-switching, each subarray is configured to radiate at broadside. The beam-switching performance of the AFR was simulated using GRASP where the radiation pattern of the feed was defined based on the simulated results of the septet array antenna. During the calculation, the focal distance is chosen to be 300 mm ($f/D = 1$) and the offset distance of the feed is 180 mm below the centre of the reflector to avoid the blockage of the reflected waves from the reflector. In this example, the feed array forms a different set of subarrays in one direction so it scans its

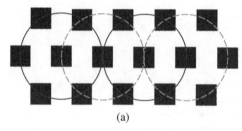

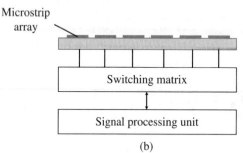

Figure 7.33 (a) Top view of the feed array. (b) The system architecture of the feed array.

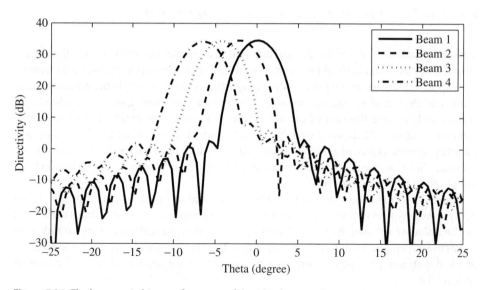

Figure 7.34 The beam-switching performance of the AFR shown in Figure 7.32.

beam in one direction. A two-dimensional beam scan can be obtained using a planar feed array of larger size.

It is noted that the feed array configuration shown in Figure 7.33 requires that the adjacent feed subarrays share the neighbouring radiating elements. This brings some design challenges to the feed network design when the number of the array element is large. To reduce the complexity of the feed switching network, the feed array can be configured to have no shared element [31], as shown in Figure 7.35. Using this approach, each subarray is designed to be fed by individual power dividers and the beamforming network only needs to switch between these seven sub-subarrays. As the trade-off, this method

Figure 7.35 The feed array used in [31], where the hexagonal array grid with 49 elements is subdivided into seven groups.

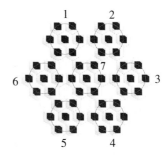

leads to increased directivity variations during beam-switching because the angle of the beams is switched with larger steps. To overcome this issue, the distance between the array elements should be reduced as much as possible, as long as the isolation between the elements is still at an acceptable level. At higher frequencies where the wavelength is small, reducing the electrical size of the subarray also means that there is limited space to design the feed network.

7.5.2 Beam-switching Performance

The beam-scanning performance of the AFR is determined by two factors: the f/D ratio of the reflector and the directivity of the feed array. To demonstrate how these two parameters can affect the performance of the AFR, a central-fed AFR with a diameter of 300 mm at 30 GHz is simulated in GRASP. Figure 7.36 shows the calculated beam-switching radiation patterns with different f/D ratios. It can be seen that the broadside directivity is almost the same, but a larger f/D ratio results in smaller gain variation (about 1.5 dB less) during the beam-switching, and the smaller f/D ratio leads to a larger beam-scanning range. With the same de-focusing distance, the AFR with smaller f/D ratio steers the beam to $-34°$, which is 7° larger than the case where $f/D = 1$.

Figure 7.37 shows the calculated radiation patterns using a feed with different directivity. In these simulations, the f/D ratio was fixed to be 1. As shown, with a higher directivity feed the beam of the AFR scans to $-27°$ and the directivity varies from 35 dB to 28.1 dB. When the feed of lower directivity is used, with the same de-focusing distance, the beam of the AFR scans to $-22°$ and the directivity varies from 32.7 to 27.8 dB. It is concluded that using lower directivity feeds results in smaller directivity variations and smaller scan angle range than using higher directivity feeds. It is also found that using a lower directivity feed array also decreases the directivity of the reflector, which is caused by the decrease of the illumination efficiency. Thus, this is a trade-off between the beam-scanning gain variation and the aperture efficiency of the reflector.

The gain variation of the AFR at different scan angles is caused by the decrease in aperture efficiency when the feed antenna is placed at the de-focused position. One approach to improving reflector illumination is to incorporate a beamforming network in the feed array [32]. By utilising the beamforming, the radiation pattern of the feed array can be shaped to obtain optimum reflector illumination and thus improve the aperture efficiency. For example, by applying aperture tapering, the beamwidth of the feed array can be increased and the sidelobes are decreased, which allows for better reflector illumination. When the beam-scanning angle is large, the illumination efficiency of the reflector

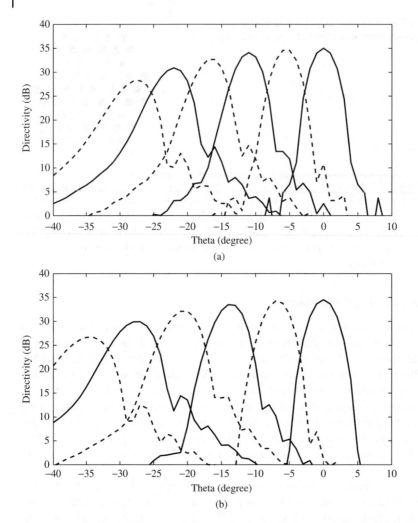

Figure 7.36 The beam-scanning radiation patterns of the AFR using different f/D ratios: (a) $f/D = 1$ and (b) $f/D = 0.8$.

can be improved by steering the beam of the feed array towards the centre of the reflector. Figure 7.38 compares the calculated radiation patterns of the AFR at $-27°$ with and without steering the beam of the feed array antenna. With the beamforming on the feed array, the directivity of the reflector is increased by more than 3 dB. Therefore, by incorporating the beamforming network in the feed array, the beam-scanning performance of the AFR is significantly improved. Although this approach increases the cost and complexity of the feed array antenna, the cost is low compared to that of the phased array antenna.

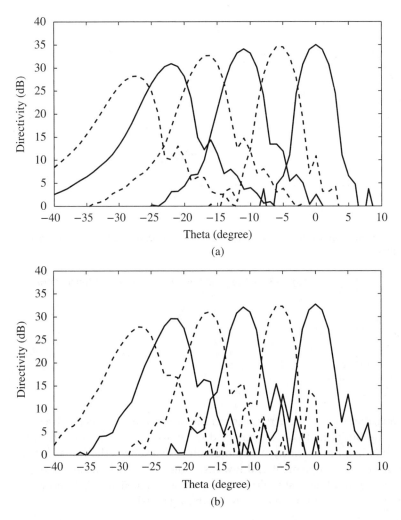

Figure 7.37 The beam-scanning radiation patterns of the AFR using a feed with (a) high directivity and (b) low directivity.

7.6 Multibeam Antennas based on BFN

7.6.1 Butler Matrix

The Butler matrix is a type of beamforming network that has been widely used in the field of beam-switching and multi-beam antenna array design. The Butler matrix has N input ports, which are isolated from each other, and N output ports. Depending on which input port is excited, the outputs of the Butler matrix produce different phases

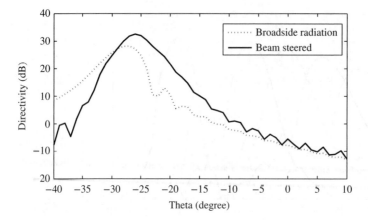

Figure 7.38 Calculated radiation patterns of the AFR at $-27°$ with and without steering the beam of the feed array antenna.

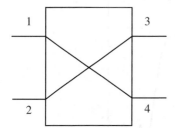

Figure 7.39 The generic version of a four-port Butler matrix.

distributions. Thus, the beam of the array antenna can be steered to a specific direction if the outputs of the Butler matrix are connected to an array antenna. If multi-beam is required, multiple input ports need to be simultaneously excited. The Butler matrix has the advantages of easy implement and low insertion loss, which make it a popular approach to designing a low-cost multi-beam smart antenna. Figure 7.39 shows the generic version of a four-port Butler matrix. Its scattering matrix is

$$B = \left(\frac{1}{\sqrt{2}}\right) \times \begin{bmatrix} 0 & 0 & 1 & -j \\ 0 & 0 & -j & 1 \\ 1 & -j & 0 & 0 \\ -j & 1 & 0 & 0 \end{bmatrix} \tag{7.21}$$

Let $Input_i$ and $Output_j$, respectively, represent the input and output of this four-port system. It can be derived that

$$\begin{bmatrix} Output_1 \\ Output_2 \\ Output_3 \\ Output_4 \end{bmatrix} = B \times \begin{bmatrix} Input_1 \\ Input_2 \\ Input_3 \\ Input_4 \end{bmatrix} \tag{7.22}$$

Solving this matrix, the following expression can be obtained

$$Output_1 = 0 \tag{7.23}$$

$$Output_2 = 0 \tag{7.24}$$

$$Output_3 = \left(\frac{I_1}{\sqrt{2}}\right) \times \angle 0^0 \tag{7.25}$$

$$Output_3 = \left(\frac{I_1}{\sqrt{2}}\right) \times \angle\left(-\frac{\pi}{2}\right) \tag{7.26}$$

The four-port Butler matrix can be realised by using the branch line coupler. As shown in Figure 7.40, each transmission line is a quarter-wavelength long at the frequency of interest and the horizontal microstrip lines have an impedance of $0.707 \times Z_0$, where Z_0 is the characteristic impedance of the microstrip line.

Several four-port Butler matrices can be connected by introducing the crossover structure to design a larger Butler matrix, e.g eight-port or 2^n-port. Figure 7.41 shows a generic version of an eight-port Butler matrix incorporated with a four-element linear array antenna. By exciting different inputs (Ports 1 to 4), the outputs (Ports 5 to 8) have equal amplitude but different phase distributions, which leads to four different beams off the broadside. Table 7.2 summarises the phase of output signals when different input ports are excited.

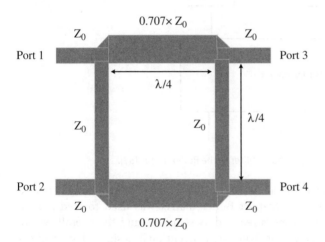

Figure 7.40 The layout of a microstrip branch line coupler.

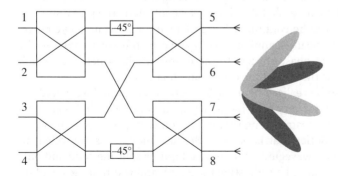

Figure 7.41 A generic version of a eight-port Butler matrix.

Table 7.2 The phase of output signal when different ports are excited.

	Output Port 1	Output Port 2	Output Port 3	Output Port 4
Input Port 1	$-45°$	$-90°$	$-135°$	$180°$
Input Port 2	$-135°$	$0°$	$135°$	$-90°$
Input Port 3	$-90°$	$135°$	$0°$	$-135°$
Input Port 4	$180°$	$-135°$	$-90°$	$-45°$

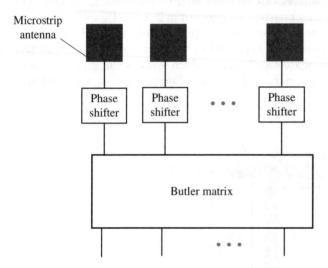

Figure 7.42 The concept of introducing a phase shifter to the Butler matrix [33].

A hybrid solution, which introduces adaptive control in the beam-switching network, was developed in [33], where an eight-port Butler matrix was used to feed a 1×8 printed patch array. The feed network is realised by striplines and the overall system has a multi-layered structure. Different from the conventional design, phase shifters were introduced in the feed network, as shown in Figure 7.42. The beam of the array is initially switched to a certain direction through choosing the corresponding port of the Butler matrix, and then it can be slightly adjusted by controlling the phase shifters. Thus, the beam of the array can be continuously steered within a certain angle range. This approach improves the beam-scanning performance of the smart antenna by introducing a small number of phase shifters to the beam-switching network, which does not significantly increase the cost of the overall antenna system.

The Butler matrix can also be realised by using a substrate-integrated waveguide as well. This is particularly useful for very high-frequency applications, e.g. millimetre frequencies, where the loss of the beamformng network becomes higher, especially when the number of ports of the Butler Matrix is large. Because the SIW is equivalent to the dielectric-filled rectangular waveguide, the required power splitting ratio and output phase difference can be achieved by adjusting the length and width of the coupling section. Figure 7.43 shows the configuration of a $90°$ SIW Butler matrix coupler.

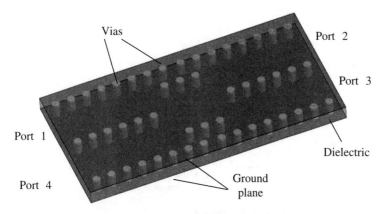

Figure 7.43 Butler matrix realised by using SIW reported in [34].

For a compact size SIW Bulter matrix, the footprint of the circuit can be reduced by using the multilayer configuration and folding the Butler matrix. In [35], a miniaturised SIW multibeam array antenna operated at 60 GHz was presented. Low-temperature co-fired ceramic (LTCC) technology was used to fabricate the folded Butler matrix and the antenna elements. The dual-CP model has 21 layers of stacked ceramic tapes. The antenna is a 4×4 array and the total circuit area is 16.5 × 14.6 mm ($3.3\lambda_{60GHz} \times 2.9\lambda_{60GHz}$), excluding the additional feeding structures

7.6.2   Rotman Lens

The Rotman lens was introduced by Rotman and Turner [36] in the 1960s. It is an RF beamformer that has N inputs and M outputs. The RF signals from the input ports (beam ports) propagate through the lens cavity and are received by the output ports (receiving ports) before transmitting to the antenna array. The positions of the beam and receiving ports as well as the transmission line lengths are calculated by using the equations of optical path-length equality so that the desired phase and amplitude distributions can be obtained. Similar to the Butler matrix, exciting corresponding input ports can produce a beam of different scan angle, and multiple beams can be obtained by simultaneously exciting multiple input ports. Figure 7.44 shows the configuration of the Rotman lens. The Rotman lens has six basic design parameters: focal angle α, focal ratio β, beam angle to ray angle ratio γ, maximum beam angle ψ_m, focal length f, and array element spacing d [37]. The β and γ are defined as

$$\beta = \frac{f_2}{f_1} \tag{7.27}$$

$$\gamma = \frac{\sin \psi}{\sin \alpha} \tag{7.28}$$

The procedure and equations to calculate the arc of the lens are described in [36]. The arc has three focal points, including two off-axis focal points and one on-axis focal point. The following equations should be satisfied to design the lens

$$F_1 P + W + N \sin \alpha = F + W_0 \tag{7.29}$$

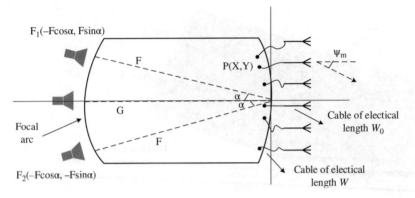

Figure 7.44 The configuration of a Rotman lens and its parameters.

$$F_2P + W - N \sin \alpha = F + W_0 \tag{7.30}$$

$$G_1P + W = G + W_0 \tag{7.31}$$

$$G_1P^2 = (G + X)^2 + Y^2 \tag{7.32}$$

where F_1P, F_2P and GP are the path lengths from the three focal points (F_1, F_1 and G) to the point P

$$F_1P^2 = F^2 + X^2 + Y^2 + 2FX \cos \alpha - 2FY \sin \alpha \tag{7.33}$$

$$F_2P^2 = F^2 + X^2 + Y^2 + 2FX \cos \alpha + 2FY \sin \alpha \tag{7.34}$$

The effects of the key design parameters on the shape and the geometric phase and amplitude errors of a Rotman lens are described in [37]. Program codes can be written by using the design equations to calculate the coordinates of the lens contours. However, for a practical design, the effects of the interference from waves reflected off sidewalls and from other ports, the mutual coupling between adjacent transmission lines, and the transmission lines transition to the parallel plate of the lens need to be considered [38]. Thus, full wave simulations using either the finite difference time domain (FDTD) or the finite element method (FEM) are used to design and optimise the performance of the Rotman lens.

The Rotman lens can be realised as a parallel-plate microwave lens, and it offers true time delay so it has a very wide bandwidth. Compared to RF phase shifters, it has low insertion loss and can be easily fabricated at low cost. Although its size is large at low frequencies, at millimetre-waves its size and weight become acceptable. Thus, it is an attractive solution for low-cost smart antennas, especially for millimetre-wave communications. A 60 GHz wide bandwidth microstrip patch antenna array fed by a microstrip Rotman lens was reported in [39]. The lens was designed on a 121 μm thick LTCC. By using a constant path delay, the phase error was reduced to less than 0.45°. This array can scan beams to ±30° in five steering steps.

Another 60 GHz Rotman lens integrated with patch antenna arrays on a multilayer liquid crystal polymer (LCP) substrate was presented in [40]. This array can also switch its beams within ±30° angle range. The lens has five beam ports, eight array ports, and eight dummy ports, as shown in Figure 7.45. The beam-switching network consists of two

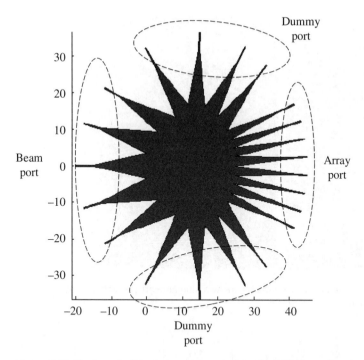

Figure 7.45 The layout of the Rotman lens calculated by [40]. Reproduced with permission of 2016 ©EurAPP.

TGS4305-FC PIN-diode switches and grounded coplanar waveguide (GCPW) transmission lines. The biasing line is printed at different layers and the beam-switching is controlled through a USB connection. Figure 7.46 shows the fabricated prototype of the beam-switching array.

Beside using a Rotman lens to feed the microstrip array antenna, other types of multibeam or beam-switching antenna system fed by a Rotman lens have been reported, including an SIW-based slotted waveguide antenna [41], an optical phased array antenna [42], and a slot-line antenna array [43].

7.6.3 Blass and Nolen Matrices

The Blass matrix is a multi-port microwave feeding network for array antennas [44–46]. Similar to the Butler matrix, it can simultaneously generate multiple beams by exciting the corresponding ports. Figure 7.47 shows the schematic of a Blass matrix. A number of feed lines in rows connect to the array antenna through another set of feed lines in columns. At each intersect point, a directional coupler is used as the crossover. The couplers are equally spaced along the transmission line. The direction of the radiated beam of the array is determined by the path difference between the input and each element. The beams of the antenna can be configured by choosing the phase and amplitude of each radiating element, which can be controlled by the feed lines and the directional couplers.

As shown in Figure 7.47, there are m inputs and n outputs, and there are no constraints on the number of the inputs and outputs. Each feed line is terminated with a matched

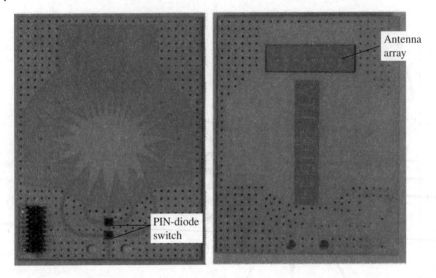

Figure 7.46 The fabricated Rotman lens with the beam-switching network presented in [40]. Reproduced with permission of 2016 ©EurAPP.

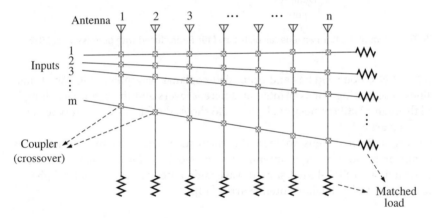

Figure 7.47 The schematic of a Blass matrix.

load for the purpose of avoiding the signal reflections. This makes the Blass matrix more lossy than other multibeam forming networks. The advantage of the Blass matrix is that it can generate arbitrary beams at arbitrary positions even with arbitrary shapes. Moreover, because of its travelling wave nature, it always has good impedance matching and the beam scans with frequency [44].

The Nolen matrix is a special case of the Blass matrix [46]. There are no terminated loads are needed in the Nolen matrix. Different from the Blass matrix, which has no constraints on the output excitations, the Nolen matrix has an orthogonal set of output excitations. Figure 7.48 shows the general form of the Nolen matrix.

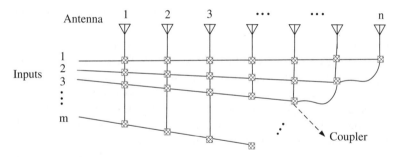

Figure 7.48 Schematic of the Nolen matrix.

The Nolen matrix has m inputs and n outputs. Unlike the Butler matrix, the number of inputs is smaller than or equal to the number of outputs [47]

$$\begin{bmatrix} Output_1 \\ Output_2 \\ Output_3 \\ . \\ . \\ . \\ Output_m \end{bmatrix} = M \times \begin{bmatrix} Input_1 \\ Input_2 \\ Input_3 \\ . \\ . \\ . \\ Input_n \end{bmatrix} \tag{7.35}$$

$$m \leq n \tag{7.36}$$

where M is a $m \times n$ matrix that linked the inputs and outputs of the Nolen matrix. Both the Blass matrix and the Nolen matrix can be realised by microstrips or SIW. A double-layer Blass matrix based on SIW technology was presented in [48]. The dimension of the Blass matrix is 4×16, representing four inputs and 16 outputs. The array antenna is a 16-element SIW slot antenna array and there are four beams at different angles within $-15°$ to $30°$. The efficiency of the SIW Blass matrix is approximately 50% and at the designed frequency the sidelobe levels of the four beams are less than 12 dB. A 4×4 planar Nolen matrix beamforming network for a Ku-band multibeam antenna was reported in [49]. The beamforming work is also implemented using SIW technology. The designed Nolen matrix consists of 3.01 dB, 4.77 dB, and 6.02 dB couplers and phase shifters with phase delay from $45°$ to $180°$. Compared to the Butler matrix, the Blass and Nolen matrices offer more degrees of freedom when designing the beamforming network. The Blass matrix can generate excitations of any distribution at the cost of lower efficiency due to its lossy network. On the other hand, an ideal Nolen matrix is lossless so it is more suitable for applications where power consumption is important.

7.7 Summary of the Chapter

In this chapter, some other types of low-cost smart antennas are presented, including the multi-beam dielectric lens, retrodirective array, pattern-reconfigurable FP resonant antenna, beam-switching antennas using the array-fed reflector, and multi-beam antennas based on passive beamforming networks. These types of antennas offer low-cost solutions to designing smart antennas instead of using the traditional phased arrays, and can achieve high-gain and multiple beams or beam-switching without the need for a large number of RF front ends and RF phase shifters. Spatial multiplexing of local element array, digital beamforming [50, 51] and hybrid adaptive antenna arrays [52] are also attractive candidates for low-cost smart antenna applications. Apart from these, there are also some other types of beam-scanning array antennas using tunable materials such as ferroelectric thin-film materials [53, 54] and liquid crystals [55–57]. These tunable materials have the disadvantage of high losses at high frequencies at the moment, but future development in materials is expected to make these tunable materials more promising for applications in low-cost smart antennas.

References

1 J. Thornton and K.C. Huang. *Modern Lens Antennas for Communications Engineering*. John Wiley & Sons, Piscataway, NJ, 2013.
2 G. Peeler and H. Coleman. Microwave stepped-index luneberg lenses. *IRE Transactions on Antennas and Propagation*, 6(2):202–207, April 1958.
3 X. Wu, G.V. Eleftheriades, and T.E. van Deventer-Perkins. Design and characterization of single- and multiple-beam mm-wave circularly polarized substrate lens antennas for wireless communications. *IEEE Transactions on Microwave Theory and Techniques*, 49(3):431–441, Mar 2001.
4 J. Ala-Laurinaho, J. Aurinsalo, A. Karttunen, M. Kaunisto, A. Lamminen, J. Nurmiharju, A.V. Räisänen, J. Säily, and P. Wainio. 2D beam-steerable integrated lens antenna system for 5G E-band access and backhaul. *IEEE Transactions on Microwave Theory and Techniques*, 64(7):2244–2255, July 2016.
5 Y. Zeng and R. Zhang. Cost-effective millimeter-wave communications with lens antenna array. *IEEE Wireless Communications*, 24(4):81–87, Aug 2017.
6 R.Y. Miyamoto and T. Itoh. Retrodirective arrays for wireless communications. *IEEE Microwave Magazine*, 3(1):71–79, Mar 2002.
7 E. Sharp and M. Diab. Van Atta reflector array. *IRE Transactions on Antennas and Propagation*, 8(4):436–438, July 1960.
8 C. Pon. Retrodirective array using the heterodyne technique. *IEEE Transactions on Antennas and Propagation*, 12(2):176–180, Mar 1964.
9 Y.C. Guo, X.W. Shi, and L. Chen. Retrodirective array technology. *Progress In Electromagnetics Research B*, 5:153–167, 2008.
10 S. Chung and K. Chang. A retrodirective microstrip antenna array. *IEEE Transactions on Antennas and Propagation*, 46(12):1802–1809, Dec 1998.
11 W. Tseng, C. Hu, and S. Chung. Planar retrodirective array reflector using dual-slot antennas. *Electronics Letters*, 34(14):1374–1376, Jul 1998.
12 A.A.M. Ali, H.B. El-Shaarawy, and H. Aubert. Millimeter-wave substrate integrated waveguide passive Van Atta reflector array. *IEEE Transactions on Antennas and Propagation*, 61(3):1465–1470, March 2013.
13 H. Zhou, W. Hong, L. Tian, X. Jiang, X.C. Zhu, M. Jiang, L. Cheng, and J.X. Zhuang. A retrodirective antenna array with polarization rotation property. *IEEE Transactions on Antennas and Propagation*, 62(8):4081–4088, Aug 2014.

14 S.L. Karode and V.F. Fusco. Frequency offset retrodirective antenna array. *Electronics Letters*, 33(16):1350–1351, Jul 1997.

15 W.E. Forsyth and W.A. Shiroma. A retrodirective antenna array using a spatially fed local oscillator. *IEEE Transactions on Antennas and Propagation*, 50(5):638–640, May 2002.

16 P.D.H. Re, S.K. Podilchak, C. Constantinides, G. Goussetis, and J. Lee. An active retrodirective antenna element for circularly polarized wireless power transmission. In *2016 IEEE Wireless Power Transfer Conference (WPTC)*, pages 1–4, May 2016.

17 P.D.H. Re, S.K. Podilchak, S. Rotenberg, G. Goussetis, and J. Lee. Retrodirective antenna array for circularly polarized wireless power transmission. In *2017 11th European Conference on Antennas and Propagation (EUCAP)*, pages 891–895, March 2017.

18 L. Chen, X.W. Shi, T.L. Zhang, C.Y. Cui, and H.J. Lin. Design of a dual-frequency retrodirective array. *IEEE Antennas and Wireless Propagation Letters*, 9:478–480, 2010.

19 D.S. Goshi, K.M.K.H. Leong, and T. Itoh. A sparse retrodirective transponder array with a time shared phase-conjugator. *IEEE Transactions on Antennas and Propagation*, 55(8):2367–2372, Aug 2007.

20 G.V. Trentini. Partially reflecting sheet arrays. *IRE Transactions on Antennas and Propagation*, 4(4):666–671, October 1956.

21 A.P. Feresidis, G. Goussetis, S. Wang, and J.C. Vardaxoglou. Artificial magnetic conductor surfaces and their application to low-profile high-gain planar antennas. *IEEE Transactions on Antennas and Propagation*, 53(1):209–215, Jan 2005.

22 S. Wang, A.P. Feresidis, G. Goussetis, and J.C. Vardaxoglou. High-gain subwavelength resonant cavity antennas based on metamaterial ground planes. *IEE Proceedings – Microwaves, Antennas and Propagation*, 153(1):1–6, Feb 2006.

23 Z. Liu. Fabry–Perot resonator antenna. *Journal of Infrared, Millimeter, and Terahertz Waves*, 31(4):391–403, Apr 2010.

24 F. Qin, S.S. Gao, Q. Luo, C.X. Mao, C. Gu, G. Wei, J. Xu, J. Li, C. Wu, K. Zheng, and S. Zheng. A simple low-cost shared-aperture dual-band dual-polarized high-gain antenna for synthetic aperture radars. *IEEE Transactions on Antennas and Propagation*, 64(7):2914–2922, July 2016.

25 R. Guzmán-Quirós, J.L. Gómez-Tornero, M. García-Vigueras, A.R. Weily, and Y.J. Guo. Novel topology of Fabry–Perot electronically steerable leaky-wave antenna. In *2012 6th European Conference on Antennas and Propagation (EUCAP)*, pages 224–228, March 2012.

26 R. Guzman-Quiros, J.L. Gomez-Tornero, A.R. Weily, and Y.J. Guo. Electronically steerable 1D Fabry–Perot leaky-wave antenna employing a tunable high impedance surface. *IEEE Transactions on Antennas and Propagation*, 60(11):5046–5055, Nov 2012.

27 A. Ourir, S.N. Burokur, and A.D. Lustrac. Phase-varying metamaterial for compact steerable directive antenna. *Electronics Letters*, 43(9):493–494, April 2007.

28 Y. Rahmat-Samii. *Reflector Antennas*, pages 668–681. Springer, New York, 2014.

29 C.A. Balanis. *Antenna Theory: Analysis and Design*. John Wiley & Sons, 3rd edition, January 2015.

30 Q. Luo, S. Gao, and L. Zhang. Wideband multilayer dual circularly-polarised antenna for array application. *Electronics Letters*, 51(25):2087–2089, 2015.

31 B. Rohrdantz, T. Jaschke, T. Reuschel, S. Radzijewski, A. Sieganschin, and A.F. Jacob. An electronically scannable reflector antenna using a planar active array feed at Ka-band. *IEEE Transactions on Microwave Theory and Techniques*, PP(99):1–12, 2017.

32 L.A. Greda and A. Dreher. Beamforming capabilities of array-fed reflector antennas. In *Proceedings of the 5th European Conference on Antennas and Propagation (EUCAP)*, pages 2852–2856, April 2011.

33 C.C. Chang, R.H. Lee, and T.Y. Shih. Design of a beam switching/steering Butler matrix for phased array system. *IEEE Transactions on Antennas and Propagation*, 58(2):367–374, Feb 2010.

34 Q.L. Yang, Y.L. Ban, J.W. Lian, Z.F.Yu, and B. Wu. SIW Butler matrix with modified hybrid coupler for slot antenna array. *IEEE Access*, 4:9561–9569, 2016.

35 Y.J. Cheng, X.Y. Bao, and Y.X. Guo. 60-GHz LTCC miniaturized substrate integrated multibeam array antenna with multiple polarizations. *IEEE Transactions on Antennas and Propagation*, 61(12):5958–5967, Dec 2013.

36 W. Rotman and R. Turner. Wide-angle microwave lens for line source applications. *IEEE Transactions on Antennas and Propagation*, 11(6):623–632, November 1963.

37 R.C. Hansen. Design trades for Rotman lenses. *IEEE Transactions on Antennas and Propagation*, 39(4):464–472, Apr 1991.

38 C.W. Penney. Rotman lens design and simulation in software [application notes]. *IEEE Microwave Magazine*, 9(6):138–149, December 2008.

39 A. Attaran, R. Rashidzadeh, and A. Kouki. 60 GHz low phase error Rotman lens combined with wideband microstrip antenna array using LTCC technology. *IEEE Transactions on Antennas and Propagation*, 64(12):5172–5180, Dec 2016.

40 J. Säily, M. Pokorný, M. Kaunisto, A. Lamminen, J. Aurinsalo, and Z. Raida. Millimetre-wave beam-switching Rotman lens antenna designs on multi-layered LCP substrates. In *2016 10th European Conference on Antennas and Propagation (EuCAP)*, pages 1–5, April 2016.

41 K. Tekkouk, M. Ettorre, L. Le Coq, and R. Sauleau. Multibeam SIW slotted waveguide antenna system fed by a compact dual-layer Rotman lens. *IEEE Transactions on Antennas and Propagation*, 64(2):504–514, Feb 2016.

42 B.A. Nia, L. Yousefi, and M. Shahabadi. Integrated optical-phased array nanoantenna system using a plasmonic rotman lens. *Journal of Lightwave Technology*, 34(9):2118–2126, May 2016.

43 J. Remez, E. Zeierman, and R. Zohar. Dual-polarized tapered slot-line antenna array fed by Rotman lens air-filled ridge-port design. *IEEE Antennas and Wireless Propagation Letters*, 8:847–851, 2009.

44 P.S. Hall and S.J. Vetterlein. Review of radio frequency beamforming techniques for scanned and multiple beam antennas. *IEE Proceedings H – Microwaves, Antennas and Propagation*, 137(5):293–303, Oct 1990.

45 F. Casini, R.V. Gatti, L. Marcaccioli, and R. Sorrentino. A novel design method for blass matrix beam-forming networks. In *2007 European Radar Conference*, pages 232–235, Oct 2007.

46 Y.M. Cheng, P. Chen, W. Hong, T. Djerafi, and K. Wu. Substrate-integrated-waveguide beamforming networks and multibeam antenna arrays for low-cost satellite and mobile systems. *IEEE Antennas and Propagation Magazine*, 53(6):18–30, Dec 2011.

47 N.J.G. Fonseca. Printed S-band 4 × 4 Nolen matrix for multiple beam antenna applications. *IEEE Transactions on Antennas and Propagation*, 57(6):1673–1678, June 2009.

48 P. Chen, W. Hong, Z. Kuai, and J. Xu. A double layer substrate integrated waveguide Blass matrix for beamforming applications. *IEEE Microwave and Wireless Components Letters*, 19(6):374–376, June 2009.

49 T. Djerafi, N.J.G. Fonseca, and K. Wu. Planar Ku-band 4 × 4 Nolen matrix in SIW technology. *IEEE Transactions on Microwave Theory and Techniques*, 58(2):259–266, Feb 2010.

50 D.S. Goshi, Y. Wang, and T. Itoh. A compact digital beamforming smile array for mobile communications. *IEEE Transactions on Microwave Theory and Techniques*, 52(12):2732–2738, Dec 2004.

51 S. Kim and Y.E. Wang. Two-dimensional planar array for digital beamforming and direction-of-arrival estimations. *IEEE Transactions on Vehicular Technology*, 58(7):3137–3144, Sept 2009.

52 X. Huang, Y.J. Guo, and J.D. Bunton. A hybrid adaptive antenna array. *IEEE Transactions on Wireless Communications*, 9(5):1770–1779, May 2010.

53 M. Rammal, L. Huitema, A. Crunteanu, D. Passerieux, D. Cros, T. Monediere, V. Madrangeas, P. Dutheil, C. Champeaux, F. Dumas-Bouchiat, P. Marchet, L. Nedelcu, L. Trupina, G. Banciu, and M. Cernea. BST thin film capacitors integrated within a frequency tunable antenna. In *2016 International Workshop on Antenna Technology (iWAT)*, pages 44–47, Feb 2016.

54 H.V. Nguyen, R. Benzerga, C. Borderon, C. Delaveaud, A. Sharaiha, R. Renoud, C. Paven, S. Pavy, K. Nadaud, and H.W. Gundel. Miniaturized and reconfigurable notch antenna based on a BST ferroelectric thin film. *Materials Research Bulletin*, 67:255–260, 2015.

55 G. Perez-Palomino, M. Barba, J.A. Encinar, R. Cahill, R. Dickie, P. Baine, and M. Bain. Design and demonstration of an electronically scanned reflectarray antenna at 100 GHz using multiresonant cells based on liquid crystals. *IEEE Transactions on Antennas and Propagation*, 63(8):3722–3727, Aug 2015.

56 B.J. Che, T. Jin, D. Erni, F.Y. Meng, Y.L.Lyu, and Q. Wu. Electrically controllable composite right/left-handed leaky-wave antenna using liquid crystals in PCB technology. *IEEE Transactions on Components, Packaging and Manufacturing Technology*, 7(8):1331–1342, Aug 2017.

57 S. Bildik, S. Dieter, C. Fritzsch, W. Menzel, and R. Jakoby. Reconfigurable folded reflectarray antenna based upon liquid crystal technology. *IEEE Transactions on Antennas and Propagation*, 63(1):122–132, Jan 2015.

Index

Low-cost Smart Antennas, First Edition. Qi Luo, Steven (Shichang) Gao, Wei Liu, and Chao Gu.
© 2019 John Wiley & Sons Ltd. Published 2019 by John Wiley & Sons Ltd.